Also by Elizabeth Marshall Thomas

Certain Poor Shepherds: A Christmas Tale

The Tribe of Tiger: Cats and Their Culture

The Hidden Life of Dogs

The Animal Wife

Reindeer Moon

Warrior Herdsmen

The Harmless People

Elizabeth Marshall Thomas

THE SOCIAL LIVES OF DOGS

The Grace of Canine Company

Illustrated by Jared Taylor Williams

Simon & Schuster
New York London Toronto Sydney Singapore

SIMON & SCHUSTER
Rockefeller Center
1230 Avenue of the Americas
New York, NY 10020

Designed by Karolina Harris
Manufactured in the United States of America
10 9 8 7 6 5 4 3 2
Library of Congress Cataloging-in-Publication Data
Thomas, Elizabeth Marshall, date.
The social lives of dogs: the grace of canine company / Elizabeth Marshall Thomas ; illustrated by Jared Taylor Williams.
p. cm.
1. Dogs—Behavior—New Hampshire—Anecdotes. 2. Dogs—Social aspects—New Hampshire—Anecdotes. 3. Dog owners—New Hampshire—Anecdotes. 4. Thomas, Elizabeth Marshall, date. I. Title.

SF433.T49 2000
636.7'0887—dc21 *99-087357*
ISBN 0-684-81026-3

This book is dedicated to those who participated in our mixed-species household:

Sundog, Misty, Pearl, Ruby, Sheilah,
Suessi, Fatima, Inookshook, Bean,
Wicket, Hobbs, Betty, and Little Sheilah

also

Lilac, Rajah, Wicca, Pae, Phyllis, Pula, Christmas,
Jeoffrey, Machka, Kochka, Carol, and Devi,

also

Lorna II, Pilgrim, Viva, Carmen, and Rima,

also

Steve, Lorna M, Stephanie, Ramsey, Bob,
Sy, Howard, Vladimir, Peter, Ele,
David, Zoë, Ariel, Margaret,
Anna, Becky, Carol, Jan,
Susan, and Don.

THE SOCIAL LIVES OF DOGS

Introduction

WHEN I was very young, my nanny was a big Newfoundland dog named Mishka, whose task was to keep me from drowning. In those days, our family took vacations on Cape Cod, and Mishka, who, like many other dogs, knew all about the command "Sit and Stay," always made me do exactly that, far back on the beach. I thus spent many summer days feeling hot, dry, and frustrated as I watched other people happily swimming, because mean old Mishka, who was vastly more strict than the grown-ups of the family, wouldn't even let me go down to the wet sand to make a sand castle.

Ever since then, I've been puzzled by the notion that dogs accept us as pack leaders. I was certainly not Mishka's pack leader. She was servile enough around the adults, always bow-

ing her head and waving her tail agreeably, but her nose and mine were of equal height if I stood up, and I can still remember the flinty look in her stern, brown eyes if I even thought about swimming. I have also been puzzled by the importance that seems to be placed on human speech. Mishka never said a word, yet she made herself perfectly clear. As a result, I learned many things from her. To be sure, I had to learn to swim from somebody else, but from her I learned how interesting and subtle someone of another species can be.

Half a century later, at the time of this writing, my husband and I still live in a mixed-species household, which we and other people share with dogs, cats, and parrots, a household that sometimes seems like a large cauldron of soup in which a mysterious force keeps churning things to the surface, always in different configurations. The many intertwined relationships often involve us, the people, but are by no means always determined by us. Sometimes we seem to be the controlling factor, and sometimes we seem to be pawns. Over the past fifteen years I have tried to look closely at the activity that was quietly going on around me, often with methods that some might find controversial. These methods were taught to me in my childhood, by Mishka, my beloved Newfoundland nanny.

Can we understand the mind of an animal? The question is hot indeed. Some say that animals have consciousness, while others vehemently insist that they don't. The rationale for this negative view escapes me since most of the research on the subject by such esteemed scientists as Marc Bekoff, Dian Fossey, Roger Fouts, Jane Goodall, Donald Griffin, Irene Pepperberg, Sue Savage-Rumbaugh, and Frans de Waal (to name a few) would certainly support the position that animals have many mental qualities similar to our own, a position which, incidentally, was shared by Charles Darwin. Observations made by other scientists such as Douglas Chadwick and George Schaller, who do not address the question of animal consciousness as such, nevertheless suggest that animals display awareness and emotion.

Some scientists, however, believe differently. Their negative view seems to spring more from theory than from field experience. Even so, the view that animals are incapable of conscious thought, or even of emotion, has acquired an aura of scientific correctness, and at the moment is the prevailing dogma, as if some very compelling evidence to the contrary was not a problem.

The negative view is bolstered by the concept that the minds of all other animals are so different from ours that we couldn't decode their mental products anyway. "If a lion could talk, we wouldn't understand him," said Ludwig Wittgenstein, voicing a sentiment that today is echoed by those who seem perfectly content to take as their guide to the animal kingdom not a biologist but a philosopher whose own mental products can only be described as murky. "Following the thoughts of a lion might be as hard for all but lion specialists as following Wittgenstein is for all but Wittgenstein specialists," says my friend Peter Schweitzer, an expert in encryption.

Wittgenstein's adherents could just as easily have turned to John Donne, much more progressive, who in 1610 wrote: "Man is a lump, where all beasts kneaded be. Wisdom makes him an Ark, where all agree."* Donne's words have been forgotten by all but the English professors, but Wittgenstein's sentiment continues to be echoed by those wishing to demonstrate the vast (as they see it) differences between animals and people. Well, of course we're different. Just as our bodies are different, our minds are surely different too. Every species is different from every other species, each one being the solution to a different set of problems posed by its environment, each the best solution that Gaia has found up to now. I'd be the first

*In "To Sir Edward Herbert at Julyers," John Donne, 1610. The verse was actually written as follows: "Man is a Lumpe, where all beasts kneaded bee. Wisdom makes him an Arke, where all agree."

to agree that if a lion could talk, we might very well not follow his reasoning on certain subjects.

For example: When doing fieldwork with wild animals in Namibia, I once watched a lion watch a sunset. He was alone on open ground near a waterhole, crouched but relaxed, propped on his elbows, and from the time the sun was about five degrees above the horizon until the last red bit of it went down, he didn't take his eyes off the spectacle. At the very last minute, he roared at it, or anyway, he roared while looking at it, not just once but four or five times, very loudly.

What did he make of what he saw? Why, on that particular evening, would he watch this familiar drama? And why did he roar? I would have loved to hear his explanation—assuming that his mind was on the sunset, as his eyes were—and I wouldn't mind a bit if I didn't understand what he said. If lions could talk, I and thousands of other people who long to know more about animals would be very happy indeed to hear what they had to say about almost anything. *Never mind that Ludwig fellow,* we'd tell them. *Nobody understands him either. Try us anyway.*

Furthermore, I believe that Wittgenstein's followers are wrong to say we'd understand nothing. Perhaps we wouldn't understand about the sun, but I'm sure that some of us would have no problem with a lion's views on a subject such as hunting. During the 1950s it was my very great privilege to live in the Kalahari Desert among the Ju/wa Bushmen,* who in those

*The many groups of Bushmen, who lack a collective term for themselves in any of their languages, were christened the San by certain anthropologists. The term *San* comes from Nama, a widely spoken language in Namibia, hence the term is now felt to be politically correct in the United States, but it isn't used in Bushmanland because it is pejorative, meaning someone who is poor, lives in the bush, has no livestock, and eats food from the ground. Although the Nama speakers sometimes use the term *San* for the Bushmen, anyone can be *san* who meets the description.

days lived solely by hunting and gathering, without domestic plants, domestic animals, fabric, or any manufactured goods. Their material resources, in other words, were the same as those of many of the creatures with whom they shared the savannah. Not surprisingly, the hunting habits of the Bushmen were similar to those of lions. The concepts seemed similar too.

Unlike Wittgenstein and his naysaying followers, who get their food from the refrigerator, neither the lions nor the Bushmen stored food—both had to hunt for it when they wanted it. Both groups needed proportionately the same amount of nourishment, both groups hunted the same kinds of prey, both groups hunted more by stealth and stalking than by chasing, and both hunted mainly by sight, making much less use of scent and sound than do many animals. Also, the behavior of the quarry was the same, the approach to the quarry was very much the same, the striking distance was almost exactly the same, and the silent expressions of restrained pleasure when the quarry was down seemed almost identical. To be sure, the Bushmen hunted only by day while the lions hunted at any time, but lions can see by starlight, so the difference was not significant.

Perhaps Wittgenstein's followers, especially those unlucky enough to live in academic confines or in suburbs or cities, would be baffled by a lion's views on hunting, but surely the Bushmen would not. Nor, for that matter, would the men who feed their families with the deer they shoot on the first day of hunting season in New Hampshire. On the contrary, such people might understand the lion quite well, and might also have something to say that he'd like to hear.

The notion that one cannot penetrate the mind of another species may seem valid to certain human beings, but it is not

---

Meanwhile, the Bushmen also found the need for a collective name, and call themselves Bushmen by choice. The term *Ju/wa* (*Ju/wasi*, plural) is the name of a certain group of Bushmen.

shared by animals. On the contrary, animals read one another's minds with surprising accuracy, and ours as well. How do they do it? At least in part, they do it by empathetic observation, by interpreting the actions of others in the light of their own knowledge, experience, and feelings. Empathy must not, of course, be confused with sympathy—empathy means that you detect a condition in somebody else and know how it feels, while sympathy means that you also have compassion. *I see that you are favoring one foot. When I do that I'm lame. Good! You must be lame. Perhaps I can catch up to you and knock you over.* That might be a cheetah's version of empathy. (Had your final thought not involved capture but sorrow—*Oh, poor thing*—you would have felt sympathy.)

If animals can fathom the mental condition of others by such a technique, then so can we. People have many of the same advantages as animals, although we may not realize it. We can see as well as most animals, at least in the daytime. We can hear as well too, within the common range of frequencies. And we can also smell, certainly if we develop our awareness of odors. We have, in other words, virtually all the sensory equipment needed to read the signals of an animal, something which human hunter-gatherers do to great effect. However, in today's climate of pseudoscience, whereby the less credit you give to animals for having recognizable thoughts and feelings, the more scientifically correct you seem, a technique such as empathetic observation raises the issue of anthropomorphism. Or it does when a person uses the technique. Nobody complains when an animal uses it.

According to the *American Heritage Dictionary of the English Language,* third edition, anthropomorphism is the "attribution of human motivation, characteristics, or behavior to things not human, such as inanimate objects, animals, or natural phenomena." Just the grouping is significant. An inanimate object, an animal, a natural phenomenon. A rock, a chimpanzee, a storm at sea. Not much difference here. When the rock glitters in the sun, or the chimpanzee bares his teeth, or the wind tosses the waves, it's only too easy to think of these things as expressing an emotion that we ourselves might feel. No wonder the naysayers are so frustrated by those of us who think there might be a similarity.

Of course, there really is such a sin as anthropomorphism. In response to an earlier book I wrote about dogs, I got a certain amount of mail *from* dogs, one of whom, a golden retriever, expressed his approval of Ronald Reagan. Now I'll grant that attributing political views to a golden retriever is probably anthropomorphizing. But attributing emotion to the grinning chimp is not, though it could be if the grin were interpreted as an expression of happiness, when actually it might be a threat. Yet a sensitive human observer would see it as a threat, because we ourselves grin with displeasure, open our mouths with aggression, and the like. What's more, the mouth would not be the only facial feature in motion. *Hey! Forget the grin! Look at that chimp's eyes!* we'd tell ourselves. *Uh oh,* we'd say, and we'd back off. We would have attributed a human motive to a nonhuman thing, and wisely so. However, such subtleties often elude the naysayers, probably because the fear of anthropomorphism is so great.

Unfortunately, the fear can blind us to genuine similarities between ourselves and other species. I once watched a television program on Laetoli and Olduvai Gorge, during which a young archaeologist speculated on two famous lines of footprints left in volcanic ash more than three million years ago by hominids. Or more likely, by a hominid and a feminid, as the prints are of somewhat different sizes. Since the prints are very close together, the anthropologist wondered aloud if their makers had their arms around each other. Then suddenly he seemed to realize what he'd just said, as if he pictured his colleagues back at the museum bristling at his sentimentality and preparing to deny him tenure. At any rate, he became embarrassed, and nervously added, "I hope I'm not anthropomorphizing."

Now I ask you! What were these creatures if not anthropomorphs? Couldn't they have cared about each other? They were certainly going in the same direction at the same time, and the lengths of their strides suggest that they were keeping pace together. Unless the feminid was slightly behind or ahead of the hominid (as she certainly might have been) they would almost have had to put their arms around each other—they were that close. Maybe they did and maybe they didn't, but was the archaeologist's speculation really so wild? More often than not, the fear of anthropomorphism is seriously misplaced.

To be sure, most animals are better at empathetic observation than we are, not only because they are more aware of small details, but also because they credit the evidence of their senses. We don't, or not nearly as much, so we miss most of what animals notice. One of the more proficient interpreters of animal behavior was Clever Hans, the German horse who appeared to understand mathematics, and who today is often mentioned in scientific works by the very people who want to prove that animals lack consciousness. If, for instance, the horse was shown something like this $\int_a^b \frac{dQ}{T} = x\text{-}Sa$ on a blackboard and was given the relevant values, he could tap the value for x with his hoof. Famous scientists came to witness the miracle of Hans, and much to the wonderment of the scientific community he solved any problem they posed.

At long last it was discovered that the horse wasn't doing any math, but was simply tapping his hoof until his questioner unwittingly signaled him to stop. Perhaps a faint sigh escaped the questioner at just the right moment, or perhaps he relaxed imperceptibly, or got a certain look in his eye. But whatever it was, Hans noted it, understood it for the request that it was, and complied.

That alone would have been quite an accomplishment. But Hans had done much more. He had accurately detected the signal to stop no matter who gave it. Many people questioned Hans, not just his owner, and each one of them must have signaled somewhat differently. The horse correctly read everybody's signal. The only time Hans appeared unable to solve a problem was when the questioner himself didn't know the answer and therefore couldn't give a signal. When the academics figured that out, the horse was exposed, and today his name is synonymous with hoax and fraud, since few seem willing to acknowledge that, as the only one in the barn who noticed the various, extremely subtle signals in the body language of his questioners, he proved himself to be a better observer than the most prominent scientists of his day.

Still, all that Hans really did was stop tapping. Our white cat, Pae, once saw me weeping, became very alert, then came bounding across the room and up into my lap, where he put his paws on my shoulders and rubbed his face on mine. He rubbed first to the left, then looked me in the eyes, then rubbed to the right and looked again. Of course, he made me feel much better. A great many other pets show similar concern, and make similar responses, often to signals far more subtle than weeping, as pet owners everywhere would agree. They read our body postures and facial expressions, they hear the tones in our voices, and they may even note emotionally generated odors, all of which are well-known signs of feelings. They understand the signals, sometimes from their own experience, and they make the empathetic leap.

The technique is more than useful, even for us. Once when doing fieldwork in Namibia, I was sitting on an observation tower watching an elephant who was drinking from a waterhole as storm clouds began to build on the horizon. The elephant was a male, about fifty years old, and he was taller than anything else in the landscape except the tower. The mopane scrub around him scarcely came up to his shoulders. If the storm came near, both he and I would have a problem, as I saw it. We were the tallest objects in the landscape, and we both would be in danger from lightning.

At first, the elephant seemed to know nothing about the storm, but when we heard a faint roll of thunder, he turned in its direction. I wondered if he might leave. He didn't. Having lived for fifty years on the savannah, he probably knew a thing or two about lightning. Although he had been moving slowly about, tossing bits of grass on his shoulders, he seemed almost indifferent until we saw the first flash of lightning, but when it came, he froze dead still.

Elephants freeze motionless for only one reason—they are listening. They all do it, even from infancy, when they want to be sure of hearing something. After the flash, all was quiet for a moment, and then came the clap of thunder. At the sound, the elephant relaxed and began to move again. But he kept his face toward the storm, and when the lightning flashed a second time, he froze once more. Four times he did this—freezing dead still when he saw the flash, moving around again as soon as he heard the thunder.

Wondering when I should leave the tower, yet not wanting to climb down right in front of this extra-large elephant, especially while something was making him nervous, I too was paying attention to the flashes and the thunder, counting the seconds between them so that I could know if the storm was moving away or coming near. To my relief, the intervals grew longer rather than shorter. The storm was moving off. No problem. I relaxed.

And so did the elephant. By the fifth flash, he was no longer freezing to listen, but had turned his rump to the distant storm and was splashing his trunk in the waterhole. Like me, he too saw no further problem.

What did this mean? Apparently the elephant had been measuring the interval between the flash and the thunder, just as I had done. When he learned that the storm was moving away, he felt safe. A more striking example of the mental ability of these remarkable animals would be hard to find, yet the only observable evidence was his freezing. An elephant standing still? So what? We often fail to make empathetic observations, and also to credit what we see.

❖

In my efforts to understand the doings of our household's animals, I began to rely on whatever measure of that ability I might possess. I'm no Clever Hans, and I can't even rival my own dogs or cats, but the method is a valid one, a good one, and also an old one, and I tried to use it as best I could. By this method, I hoped to get some notion of what our animals were thinking, to get glimpses of their worldviews, and to understand their social longings. And at last this began to seem helpful in explaining our big, churning cauldron of a household, our cluster of dogs, cats, parrots, and people. This book contains what I found.

1

OUR group of mixed species began soon after my husband and I moved to New Hampshire from Virginia. Three dogs came with us, all very old, the only dogs who remained alive from a once much larger pack that, unlike most groups to which dogs belong, contained no nondog members. In the order by which these dogs would have presented themselves, Dog One was Suessi, a powerful white and gray husky, the only male. Dog Two was a sensitive, heavyset female who looked something like a beagle. She had belonged to our daughter, and was named Fatima for a much-loved teacher. Dog Three was Inookshook, Fatima's half-sister, a mild-mannered, very gentle dog who looked like a husky, with red-gold hair and pale blue eyes.

Strictly speaking, Fatima and Inookshook were not entirely dogs because their mother was a dingo. However, dingoes are almost dogs, and Fatima and Inookshook passed as dogs, so that is what I called them except in times of need, when I had five dogs and a dingo but lived where zoning requirements allowed me only four dogs, and it became necessary to omit the dingo from my communications with the authorities, and also to count Inookshook and Fatima as half-dogs so that together they made one.

By the time we moved to New Hampshire, all three dogs had varying degrees of arthritis. Fatima was diabetic, and Suessi and Inookshook suffered from a canine form of Alzheimer's disease which made them vague and vacant—they would look at me and other people as if they didn't know just what we were, and couldn't imagine how we came to be there. But if people were not important to them, their group was. They slept together as they had done all their lives, wherever we lived, lying calmly on their blankets but nevertheless touching one another, fur to fur.

At dawn, they followed Inookshook outdoors and waited while she, as dog custom demands of the lowest-ranking pack member, carefully chose the spot which, that morning, the three dogs would mark. As in days gone by she would squat while the other two dogs, in ascending rank, waited their turns beside her. When all three had emptied their bladders on the same spot, they would slowly explore the fields, Suessi and Fatima together as a pair, Inookshook far behind the others as a loner, all of them nosing around to learn what the wild animals, especially the coyotes, had been doing during the night. If a coyote had marked, the dogs would overmark his stain to point out, in their elderly, faltering manner, that they, the dogs, were the actual owners of the field, and that the coyotes should stay in the woods.

That done, they would spend the rest of the day lying on a hilltop side by side, their rumps to the woods where the coyotes lived, their eyes on the road where, from time to time, a dog would appear in the company of a jogger. So the hilltop was like a farmhouse porch and our dogs were like three elderly country people in their rocking chairs. Retired dogs, like retired people, still want to know what's doing with their neighbors.

In the evening, from the shed, we would hear the dog-door click three times. One behind the other, the old dogs would slowly come in for dinner, first the rickety Suessi, still the leader for all his infirmities, then fat little Fatima, the waddling elder sister, then tall, slow, Inookshook, the stiffly graceful younger sister. And thus we lived for their declining years, my husband and I preoccupied with our affairs and the three dogs preoccupied with theirs—a peaceful, well-ordered existence in which each of us knew our place and our duties vis-à-vis the others.

In December of that year, my mother was scheduled for surgery. I went to her home in Cambridge, Massachusetts, to be with her. As I came up her walk, I saw a large white dog lying in a corner of her porch. I climbed the steps and stood beside him. He looked away, as if he wished he couldn't see me.

I reached slowly toward him but, unhappy and unsure, he shrank away from my hand as far as he could without standing up. His body might have been glued to the floorboards. And so, stepping back, I looked him over. He was a male, about seven months old. With his long muzzle and large, upright ears, he looked at first glance like a white German shepherd. But he wasn't quite as lanky as a German shepherd. Rather, he was square-bodied, more like a yellow Lab. Nose-tip to tail-tip, he was about four and a half feet long, and he seemed to weigh between sixty and seventy pounds. His fur was short and white, and was mostly guard hairs, with very little undercoat. His skin was pink; his ears were also pink where the sunlight shone through, but his eyes were dark. So was his nose, fading to pink where it entered his fur. In short, he was handsome. I thought he would stand up.

Strangely though, however long I smiled my welcome, he

simply hunkered low and didn't move. Wondering if he'd been hurt, I tried to examine him, but saw no blood or bruises. I did, however, notice a long surgical scar up his left foreleg and over his shoulder. He'd had an operation. Perhaps he'd been hit by a car. His hair, which had been shaved for the surgery, was still growing back. Otherwise he seemed very beautiful and clean, in excellent physical condition. Although he had no collar, no tag, no identification of any kind, someone seemed to have cared about him.

Because I was concerned about my mother—we were leaving for the hospital—I couldn't do anything for the dog except to call the dog officer at the local humane society. I couldn't even invite the dog in, since no one would be there to answer the door when the dog officer came. So I left him on the porch. When I returned alone very late that night the dog was still there.

It was cold, but, because I had become even more preoccupied with my mother and not at all able to take responsibility for a stray dog, I again left him outside, hoping that he'd travel on or that, somehow, his owners would find him. Before going to bed I called the dog officer again, but didn't reach him, so I left a message on his answering machine. But in the morning the dog was still lying on the porch.

I left a new message for the dog officer before returning to the hospital, but the dog was still on the porch that night when I came back again. By then, the temperature had fallen seriously, turning uncomfortable weather into dangerous weather. Without shelter, the dog faced hypothermia. Also, he hadn't had food or water for at least two days.

I sat down on the step beside this white dog. Trembling hard, he glanced sideways at me. I talked to him calmly and quietly for a while, my head partly turned in his direction but my eyes averted. At last he looked straight at me, ears low. I stood up slowly. He did the same, the first time I'd seen him on his feet, and reluctantly he let me touch him very gently, although he shrank away from my hand. After a while I unlocked the front door and went in, leaving it open behind me. An icy wind followed. Soon, the dog did too.

As he stood alone in the hallway, uncertain and uncomfortable, I searched in the kitchen for something I could feed him, and found a bag of dog food left uneaten by my mother's late but much beloved black Labrador, Micah. In solemn silence, the white dog ate dear Micah's food. When he finished, he looked at me steadily for a moment and then, suddenly, surprisingly, his face lit up and he frisked very briefly, raising both front feet together in a little leap. For just that instant, he seemed happy. Then he got hold of himself again, had a long drink of water, and curled up on a blanket I had placed near a radiator, where he soon fell asleep. Perhaps he had been too cold to sleep while he was on the porch, or perhaps he had been keeping a vigil for his people. In the warm house, with his hunger satisfied, he could no longer stay awake.

In the morning before departing for the hospital, I left him in the house with food, water, and a blanket, and from my mother's hospital room began an intensive search for his owners. Before the day was over I had notified all the local newspapers and radio stations and had answered all the lost dog ads, each and every one, even the ads for dogs whose descriptions in no way matched the dog on the doorstep. By evening, I had called the dog officer of every community within a radius of thirty miles because I had known two huskies, Suessi's parents, who would travel that distance. I had also called everyone I knew who might adopt the dog if the owners could not be found. And by the time my mother was discharged on the third day, I had made up dozens of fliers which I tacked to trees or posted on the bulletin boards of supermarkets and other stores. In addition, I also walked the white dog on a leash, hoping he could lead me to his home, and I drove him around in the car to see if any neighborhood seemed familiar to him. But none did.

To find his owners began to seem hopeless. I wondered if the dog had been stolen from far away, then transported to my mother's neighborhood and released there. I considered bringing him to one of the local humane societies even though I knew that after a few days he would be put to death in the very likely event that no one came along to adopt him. But I closed my mind to this possibility because at the time, although it shames me to admit it, I didn't want a dog like him. Our three elderly dogs in their well-ordered group would never have welcomed a pup or an outsider. Nor would my husband, who didn't want another animal of any description. And even though I myself was always open to another dog, I hadn't planned to get one until the three elderly dogs were no longer living.

And when that time came, I planned to get a capable, adult dog whom I could learn from—a sophisticated dog who had been well educated by other dogs. I wanted a dingo, perhaps, or an Indian dog from northern Canada or a pariah dog from a Third World village. I certainly didn't want a purebred American dog, or even a mix of purebred dogs, which was what the white stray seemed to be.

The notion of breeding dogs to a standard of appearance has always seemed peculiar to me. The important features of a dog are his brains and his persona, so that some of the best members of the dog family are, say, ordinary working sheepdogs in the backcountry of Australia, or ordinary village dogs in Latin America, Africa, and Asia, or, for that matter, ordinary dingoes or coyotes or jackals or wolves. So I didn't want a purebred dog, or even a cross between two purebred dogs. I knew that any pedigreed strains in the white dog's makeup were not his fault by any means, but I felt sure the strains were there, and I didn't want him.

Yet after four days in his company, I began to feel very moved by the quiet, forsaken youngster. Like a professional soldier he seemed alert yet disciplined, quietly ready at all times to do anything that his superiors might command. I'd catch him watching me out of the corner of his eye, asking for nothing, assuming nothing, but even so, surely wondering what his fate at my hands would be. I couldn't help but feel his isolation, or admire his reserve, his pride, his delicacy, and his intelligence, and soon I began to ask myself what, after all, is really so wrong with a few purebred strains? I continued to search for his owners, of course, but with decreasing dedication, until at last I dreaded to hear the phone ring for fear I had found them.

My husband, Steve, was still not ready to take on another pet of any description, but by then I was hopelessly attached to this white dog, and he to me. His pale fur and his chilled, obscure persona reminded me of a certain sundog, a hazy, frosty image of the sun that Steve and I once saw below the real sun on an icy winter day. And against Steve's wishes I, Elizabeth, took the white dog, Sundog, to be my dog, to love him and to cherish him, for better or for worse, in sickness or in health, for as long as we both should live. So too had he, Sundog, taken me, Elizabeth, to be his lawful person, for better or for worse, whatever might befall us, until death would us part. Somehow, in spite of his long vigil on the porch, in spite of all my efforts to find his owners, and before either of us fully understood what was happening between us, the white dog and I became one.

2

WHEN Sundog first came, our household contained two distinct groups. The members of one group were dogs—the elderly Suessi, Fatima, and Inookshook—and the members of the other group were people—my husband and me. We had no cats or other kinds of pets, as the husky and the dingo-dog crosses would have shredded them. As groups we kept more or less apart, each friendly with the other but separate nevertheless as we minded our own business and involved ourselves only with the affairs of our own kind. When Steve and I took walks, for instance, it didn't occur to these dogs to come along. They would have been welcome to join us, of course, and because of the dog-door and the fact that they were free to choose their own activities, they could have joined us if they wished.

They never did. Like people who take very little notice of animals, they viewed our doings incuriously. They took walks often enough, but only with one another and only to places of their own choosing, and whenever Steve and I set off, instead of following us as other dogs might, they merely watched without much interest until we were out of sight. In the same spirit, they paid very little attention to human visitors. They never barked when a car drove up or bothered to investigate whoever got out, rightly assuming that we, the human beings, would keep the newcomers under control, and that the reason for the visit would not be interesting or important in a dog's eyes. They also found no cause for excitement when, one night, a burglar entered our home and hence they did nothing to stop him. Invasion by a strange dog would have been an entirely different matter, but the burglar was a person and dogs weren't involved. Then as always the old dogs had minded their own business—dog business—just as Steve and I always minded ours. After all, if a dog passed by on the road, it didn't occur to Steve and me to wonder if our domain was under invasion. Nor did we concern ourselves with the coyotes. As far as we were concerned, they were welcome to come out of the woods if they wanted to. (If a person were to emerge from the woods, however, we might have taken him for a poacher and called the game warden.) We all were very species-oriented in those days.

When my mother was well enough for me to leave her, Steve came down from New Hampshire to get me, and drove me and Sundog home. Perhaps Steve hadn't wanted another dog, but he was very understanding about this one, and wondered aloud how our old dogs would behave toward him. It seemed possible that because the old dogs had once belonged to a much larger pack, they might be glad to see their number increase again, even if only by one. But our hopes were in vain. We might as well have expected that three elderly residents of a retirement community would invite a lonely teenage stranger into their rooms from the streets.

At first, Sundog saw things differently. He sensed no distinction between himself and the old dogs. He was a dog—a forsaken, lonely dog—and they were dogs too. As we came up the driveway, he was overjoyed to see them standing on the lawn, and when he got out of the car he ran to them offering a young dog's friendly greeting—his knees and elbows bent, his chin high, his neck stretched, his head and ears politely low, and the tip of his tail repressed but nevertheless waving—waiting for their inspection and their welcome, which would release him to wag not only his tail but also his entire body with all the enthusiasm in his young heart.

But the three old dogs scarcely bothered to find out who he was, giving him little sign of recognition and no welcome. He might not have been a dog, for all of them. He persisted, kissing the corners of their lips to emphasize that he was young and eager to honor them as his elders, but they bared their teeth, withdrew their faces, and walked away from him.

Sundog was crushed. As his hopes faded, he made himself small, youthfully curling up near us with lowered ears, humbly waiting to see where his place would be in this new household. That night at our invitation he followed us to our bedroom although he would rather have slept with the old dogs in the entrance room between the shed and the kitchen where they, incontinent because of age and Alzheimer's, spent the night together lying on waterproofed mats.

The next morning, Sundog amazed us by enthusiastically rushing to the closed door of the entrance room and urinating on it. Horrified to see him lift his leg inside the house, I shouted at him. Equally horrified, he stopped his urine in midstream and turned his head, dismayed, to look at me over his shoulder. Seeing my expression, he immediately realized that he was doing something terribly wrong. He then seemed overcome with shame and, lowering his leg, he crept away from the door. I said nothing further because even I, human being that I am, suddenly understood his motive. Due to the elderly dogs' incontinence, the scent of urine wafted from the entrance room, an odor so strong that even I could smell it. Obviously Sundog had simply taken for granted that everything that the old dogs did was right. If they urinated in the entrance room, then so must he. *The elders show the way. If you want their acceptance, do as they do,* says an important dog law, which most young dogs obey.

The full significance of Sundog's leg-lifting was revealed later. He never would have done such a thing carelessly, and unlike most of the other dogs we'd had, he never had an accident inside the house. He wouldn't even move his bowels on the lawn. For many years, until arthritis prevented him, he always journeyed far away, to the privacy of a distant field. Hence, to have urinated on the kitchen door must have been, for him, an act of great meaning.

As we soon observed, his house-training had been early, rigid, and very severe. Once, when an emergency forced us to board him in an unfamiliar kennel where he was kept in an enclosed concrete pen that must have seemed to him like the interior of a building, he held his urine until we came to collect him two days later. Going home in the car he cried, but we didn't know why, and when he reached our lawn, he descended from the car very slowly, then lifted a hind foot a few inches from the ground and, eyes shut and ears low with discomfort, he urinated painfully and steadily for almost two minutes. We then understood what he must have been going through. Yet what he would not do even to relieve his own intense suffering, he had done with enthusiasm in hopes of acceptance by our other dogs. This showed, I think, the strength of his desire to belong among them.

My husband and I soon realized that we had in our household a truly extraordinary being. Sundog was extremely bright. One day, I tested him with a test that was allegedly designed by behaviorists or psychologists to evaluate the intelligence of dogs. Suspecting it to be derived from some kind of object-permanence test such as those devised by Piaget for children, but believing that, in the case of dogs, it actually proves very little, I give the test anyway, because most dogs like it. It involves a dog biscuit which, in the presence of the dog, I place on the floor and cover with a towel. To get the biscuit, the dog must supposedly remove the towel. If he gets the biscuit, he passes the test. That's all there is to it. Allegedly it proves that the successful dog realizes that the biscuit is present even if no longer in view. (This rationale is unclear to me, since the dog, of course, can smell the biscuit.) Still, Sundog loved this test, at which he excelled, nosing the towel aside with lightning speed and grabbing the biscuit, thus passing the test again and again as long as the supply of biscuits lasted and demonstrating in a manner that would satisfy even the fussiest behaviorist that although he couldn't at the moment see any biscuit, he certainly knew what was under the towel. To me, it was a joy to see a dog solve a problem quickly and easily, and whenever I gave the towel-biscuit test to Sundog, the splendid creature did just that.

The test gave him a taste for dog biscuits, causing him to initiate a custom that we kept from that time on. When one of us would remove a biscuit from its box, his face would take on an alert expression. Presently he began to use the expression before anyone went to the box. Seeing his face, we would then go to the box and get him a biscuit. He then began to display this expression at a certain time in the afternoon. We took to giving him a biscuit at the same time every afternoon. One afternoon, however, I was sitting on a bench, talking with Steve, my husband, about something important. Sundog began to look at Steve with raised ears and bright eyes, but Steve went on talking. Sundog barked. Steve still paid no attention. Sundog jumped up beside me and stood in front of me so that Steve suddenly found himself looking into Sundog's face, on which he read the fact that the time had come to get up and get a biscuit.

If Sundog had no difficulty in communicating his wishes to us, he also understood our wishes, in part because he understood a great many words of English. He quickly learned our names and his new name, of course, and the usual commands—*come, go, stop, sit, stay, no, okay, heel, down, lie down,* and so forth—but after he had been with us for a while we realized that he knew many more words as well. Those with dog relevance such as *walk, road, town, woods, car, bed, dog, cat, door, dog-door, downstairs, upstairs, indoors, outdoors, kitchen,* and *veterinarian* were plain enough to him, to say nothing of *breakfast, dinner, water, cookie, treat, popcorn, dish,* and *dog biscuit,* but I often wondered if he didn't understand many others too. If we said to him, "Sundog, would you please step off the rug? I need to move it," or "Go ahead, I'll meet you by the car," for instance, he'd do it, calmly and proudly with dignity and quiet grace, like a soldier.

His honor was more important to him than anything else—a rare quality in an ordinary pet dog—so that he complied with every request and obeyed every command. He also guessed correctly what was wanted of him without being told—for instance, he never once in the many years since we found him raided a garbage pail or took food off a kitchen counter or a coffee table. We could leave him alone in the car with groceries that included cheese or meat or dog treats. Sundog touched nothing. He wouldn't even touch food that fell on the floor unless invited. No other dogs of ours had ever been so hard on themselves. When left alone they persuaded themselves that it was okay to raid the trash or whisk away a tasty morsel left too near the edge of a kitchen counter. An outdoor party we once gave, a cookout which my uninvited dogs somehow managed to attend, was termed "Paleolithic" by a guest. "This isn't a cookout," said the guest as the pack of greedy dogs stormed around us, shouldering one another and snuffling over our plates, "this is a hunting camp in the Upper Paleolithic!"

Nor would our other dogs bother to travel all the way to a field just to relieve themselves as Sundog did. Far otherwise. In cold or rainy weather at least one of them would, unless we objected loudly and quickly, relieve herself right on the threshold. And sometimes these dogs, when feeling sick during the night, instead of coming to us to ask for help, misused the dining room rug. But even after great age and arthritis incapacitated him, Sundog made no such errors.

3

When Sundog first came, his interest in us and his desire to understand what we wanted made him seem refreshingly different from our older dogs, who had more or less ignored us. Sundog was going to be a joy—we could tell. But very soon we discovered his dark side—an unusual, steely stillness that seemed better suited to a much older dog than to one so very young. To me, this too spoke of his early training, which like his housebreaking must have been mortally severe. Surely as a result of some very heavy training that could only have taken place in his extreme youth (since he was just six or seven months old when we found him), he had certain deep fears that would have him shaking violently, his eyes stretched wide, his ears folded flat, his teeth

chattering. He was, for instance, afraid of children. When a child appeared, he'd vanish mysteriously behind a door or under a table. Once or twice a child sought him out to pat him. Thus cornered, he barked sharply and snapped, upsetting and scaring the child. Sometimes if wakened by touch while asleep, he'd leap up with a roar, terrified and instantly ready to defend himself.

His worst fears involved furniture. The first time we invited him onto furniture he averted his eyes, raised his chin, shut his mouth tightly, and lowered his ears to make his head and face sleek. He also trembled violently. We saw that he was experiencing a terrible conflict—he trembled because we, his family and his saviors, were asking him to do something he apparently believed was very wrong.

We had asked him onto the couch as a way of relaxing him. Because when he was in our company he often sat stiffly, as if at attention, ready to leap up when we stood, we had only wanted to show him that we were more casual than his former owners, and that in our house dogs were welcome anywhere, even on our furniture. But when our friendly advances upset him, we of course dropped the subject for the time being, and didn't press him.

But he wanted to join us, I could tell. He gazed with open longing at the scene of us on the couch making a space for him, as if he were witnessing an unattainable dream. I thought perhaps that he was envisioning himself on the couch with us. But soon his early, violent training overrode his wish. He stood up and crept away behind a chair where he couldn't see us or the couch, and thus removed himself from the conflicting demands of what he wanted, what we wanted, and what he surely had been severely trained never, ever to do.

All his life he held deep convictions about what furniture he would and would not use. With a few extraordinary exceptions, he never got up on any bed for any reason. Nor would he use the living room furniture. However, he would gladly sleep on a couch on the porch, on a small couch in my husband's office, and on a couch in my office. He'd stay there through thick and thin too, even if we also wanted to sit there and tried to make him move. Without looking at us he might squeeze forward a little, yielding a few inches of space, but if we persisted, he would flatten his ears, half shut his eyes, hunker low on the couch, and passively resist with all his might. *I know what you want,* his manner said. *But I won't do it.*

When we first met him, his dreadful fear of furniture found its climax in his fear of people's beds. Early in our relationship with him we invited him onto our bed, but the invitation distressed him so deeply that he had to leave the room. Once we lifted him onto the bed to show him that it was really okay for him to be there, but he began to tremble so violently that his teeth chattered and, looking at us with both terror and apology, he slid off the bed as unobtrusively as he could. A moment later we looked around for him and couldn't find him. He had hidden himself in a closet.

Only once did I ever find Sundog on a bed. I had entered the bedroom from the shower, ready for the night, and was astonished to see him lying full length in my place on the bed, just as I would do, with his head on the pillow, back to back with my sleeping husband. Sundog raised his head slightly and stared at me as I came in, his ears low, his mouth set, as if waiting for me to disapprove. *Now she's going to chase me off,* his manner said. But I was so glad to see that he'd at last relaxed his standards that I just stood still. His eyes met mine and we watched each other. He was waiting for me to speak. But instead, just for fun, I lay down on the floor on his bed.

Sundog was astounded. His eyes flew wide, his mouth opened and his ears lifted as he jerked himself upward to stare. *What on earth is she doing?* Again our eyes met. And suddenly, he got it! A dog joke! He seemed delighted. His face brightened with a dog smile and he thumped his tail happily. Then, his tail waving with pleasure, he jumped to his feet and off the bed, and we changed places. I invited him to join us on the bed, but he didn't. Rather, he lay down on his own bed and left me to mine, but he seemed very content.

That he had been so rigidly trained increased the mystery of his presence on my mother's porch. So did the surgical scar that ran from his elbow to his shoulder. Obviously, his former owners had spent much time and effort to educate him and a great deal of money to heal him. Many years later we learned how much they must have spent when our veterinarian discovered from an X-ray that Sundog's bones had been shattered and pinned back together by a state-of-the-art technique, a very costly operation that at the time could only have been performed at one of the most sophisticated animal hospitals in the country, such as Angel Memorial in Boston, or at the veterinary college at Tufts or Cornell. But even though we didn't know this at first, it was obvious enough from Sundog's training and surgery that his former owners had invested heavily in him. Why, then, hadn't they tried harder to find him?

Why hadn't they at least run advertisements, or looked for my advertisements, or checked with the local humane societies or the dog wardens in the various towns? I'd had plenty of experience with lost dogs—mine and those of others—and I had never failed to find them. Sometimes I succeeded only after prolonged and strenuous effort, but I'd always succeeded. After all, any lost dog is somewhere. No dog just vanishes from the face of the earth, and almost every lost dog can be found sooner or later, assuming that the dog is living and the searchers persist.

One day we found a possible explanation. Shortly after Sundog came to live with us, he accompanied me when I took the diabetic dingo-dog Fatima to the vet. Sundog was hostile to veterinarians, and once had bitten a veterinarian, so when Fatima and I headed off toward the clinic he showed that he wanted to stay in the car.

I let him. After all, he was extraordinarily self-sufficient, quite capable of making his own decisions. When we came to know him better we learned that before settling himself for a ride in a car he would stand briefly on the seat to test the temperature. If he found the car too hot, he'd climb out again, giving us a glance to tell us that he'd ride with us another day. If, on the other hand, he chose to come, we'd park in the shade and leave the windows open. On the rare occasions that he misjudged the heat he would simply leave the car through the open window, find himself a shady place, and wait for us there. No trouble ever came from his freedom.

But even at the time, I knew he'd be safe if left in the car at the clinic, so when Fatima and I got out, I told Sundog to stay. I wanted him to know that he needn't accompany us. He settled himself on the seat and was there when we returned. But when we went home and Fatima and I got out of the car, Sundog didn't move.

This was in April, four months after he had come to live with us. We didn't know him then as well as we came to know him later. Although his staying behind struck me as odd, I casually assumed that, like other dogs I'd known, he enjoyed sitting in a car and would get out when he felt like it. So I left the door open. But when I reached the house the situation began to nag at me, and I went back. Sundog hadn't moved. Suddenly it occurred to me that he was waiting for something. Properly trained dogs are supposed to keep on doing what they were told to do until the trainer tells them to stop.

Sundog looked at me. I looked at him. "Okay!" I said. He stood right up and came out, and as his long, white body moved down to the grass it struck me that someone might have ordered him to sit and stay on my mother's porch, and never came back to release him.

The more I thought about this, the more likely it seemed. My mother had a far-reaching reputation for her kind heart and her love of animals, and Sundog wasn't the first animal to find himself on her doorstep. Unknown persons had left kittens in paper bags and shoe boxes. My mother had taken them in. Maybe someone had thought to try to give her Sundog.

But what did this say about him? For three days and two nights as he waited—frightened, lonely, starving, thirsty, and almost frozen from the winter wind that bit through his thin fur—he had slowly been declining toward hypothermia and death. Yet his soldier's honor had kept him in place, enduring every terror and discomfort, forcing every ounce of his being to do what he had been asked to do, never weakening, never once breaking his pledge to those who left him there. Who were they—they who accepted the trust of so honorable a dog, and then so deeply betrayed him?

4

WHEN Sundog saw that he had no choice but to accept the decision of the elderly dogs to exclude him, he joined our group, and thus became the first member of our household to cross species lines. Perhaps we were not his first choice, but he took us. At least with us he could assume a higher rank than he would have held otherwise. He would have held Place Four in the dog group, and while he certainly knew that in our group we, the human beings, outranked him, we were not dogs and therefore didn't outrank him as definitely or as poignantly as other dogs would. Also, if his status was less than ours, we didn't keep reminding him of it, as the other dogs would have done. Perhaps he couldn't be First in the Household, but he could certainly be First Dog in the Human Group.

With that, Sundog committed himself to us, the people. He eventually demonstrated such a complete understanding of human behavior that if I believed in reincarnation, I'd have been convinced that in an earlier life he'd been a person. Or if I believed in enchantment, I'd have taken him for an enchanted person who, like the Frog Prince, had been imprisoned in an animal's body, awaiting the kiss of a princess's lips to return him to his human state. That was how well he understood human behavior and mannerisms.

One of his first acts was to bond with Steve. His bonds with me were never broken, and he was always extremely loving and kind to me (as I was, I hope, to him), but no sooner had he noted that Steve was Person One in our family than Steve became his human of choice and he set about courting him. Wherever Steve was, there Sundog was too. Perhaps Steve hadn't wanted another dog at first, but after Sundog made up his mind to join Steve, Steve had little to say about the arrangement, and very soon he and Sundog were inseparable. Side by side they walked in the fields, and side by side they drove in the car to town.

One day it came to us that Sundog might be emulating human mannerisms. In the car with Steve he'd sit with his rump on the back seat, human-style, and his hind feet squarely on the floor. The position enabled him to lean forward and place his front feet, like hands, between the bucket seats. This put his head beside Steve's. Together, they could face the road ahead, and so they remained while the car was in motion. When the car stopped and Steve got out to do his errands, Sundog would take Steve's place behind the wheel, a spot which many dogs realize is Place One, or the most significant place inside a vehicle. There he would await Steve's return, when he'd move over to the other front seat, or take his humanlike position on the back seat again.

Part of Sundog's display of human mannerisms was his attention to human speech. He fully realized that speech meant everything to us, and hence he listened to us carefully. How much so was revealed by, of all things, a squirrel in a bird feeder. I would look out the window, see the squirrel, say something like, "Oh, the squirrel is back," then go outside and chase it away. Soon, Sundog was coming with me to chase it, and before long, all I needed to say was "Squirrel" and he'd rush out the dog-door and chase the squirrel away by himself. Winter came, and long nights, and on one of those nights I noticed a raccoon in the bird feeder. I was in my nightgown and didn't want to go out in the snow, so I said to Sundog, "There's a raccoon in the bird feeder." But Sundog didn't know the word "raccoon." He looked at me, wondering. I had a bright idea, or so I thought, and said, with enthusiasm, "Squirrel! Squirrel!"

Sundog leaped to his feet and rushed for the dog-door, his toenails scrabbling on the linoleum. I looked out the window, waiting to hear the dog-door slam and then to see Sundog charge around the house like a white knight defending his castle, but just as I was wondering why the raccoon seemed unconcerned, I heard Sundog's toenails slowly clicking on the floor as he came back toward the kitchen. He hadn't left the house. As he entered the kitchen, he gave me a look of deep disappointment and flung himself down on his bed. He was hurt, or at least disgusted. What had happened?

I'll never be sure, of course, but it came to me that he thought I had deceived him. Squirrels are creatures of the daylight. At that time of night, there couldn't be a squirrel. Sundog would have known all about things like that. And whatever went wrong, whatever thought passed through his mind that made him reverse direction, that thought had come to him while he was still in the entrance room on his way to the dog-door. As the entrance room was on the far side of the house from the bird feeder, he could not have known about the raccoon. I apologized very much, then tried to bring him to the window so that he could see for himself that there actually was an animal, but, as if I'd played a practical joke on him, as if I'd made a fool of him and taken advantage of his willing nature, he wouldn't come.

After a few months, I noticed that Steve was paying a great deal of attention to Sundog. He didn't want to be apart from Sundog any more than Sundog wanted to be apart from him. They went to bed at the same time, got up at the same time, and walked in the same way to the same places. Obviously, they respected each other. Sundog would look up when he heard Steve speaking, and Steve would go to the window when he heard Sundog bark. Each would leave space for the other on the couch in Steve's office, and neither interfered with the other's possessions, so that Sundog never chewed Steve's gloves or shoes and Steve tolerated even the filthiest bones on the floor of his office. They took each other's part in every question, and took each other's friends and enemies as their own. On trips to town, they would pass other dogs whom Sundog vilified, and Steve would swear at them too, although, before Sundog joined us, Steve had passed these same dogs daily without noticing them. Sundog always chose Steve over me if we took different directions, and one day I asked Steve if he would choose Sundog over me if the question arose. Steve gave me a deep, uneasy look and thought for a very long time before speaking. "Don't ask me that," he finally said.

In time, Sundog's human mannerisms began to seem uncanny. For instance, he and Steve would share ice cream cones. Steve would take a small bite and pass the cone to Sundog, who would also take a small bite, and so on down to the tip of the cone. Dogs, of course, never share food with one another in this manner, although they sometimes eat together from the same food source. But when they do, they both gobble at the same time, avoiding only the food that rests in a small, inviolate circle around the other dog's mouth. Taking turns is never an option. So Sundog's behavior was a true departure from that of dogs, and was entirely human.

Needless to say, nothing delights a person more than to be

copied by an animal. They lack our skills, but they try anyway. This is why circus elephants are made to wear tutus and to place their gray behinds on stools, and why we want dogs to walk on their hind legs and parrots to speak our languages. Steve and I were predictably enthralled by Sundog's sharing and we shamelessly encouraged him.

Thus we were delighted to find that ice cream was not the only food he'd share in this very human manner. One evening, somewhat later, Steve was sitting at the kitchen table eating popcorn from a small bowl. Soon, despite his aversion to furniture, Sundog quietly clambered into the chair next to Steve's, seated himself, and glanced at the bowl. Steve understood, and handed a popped kernel to Sundog, who took it delicately. Then Steve ate a kernel while Sundog watched, and Sundog ate a kernel while Steve watched, and so they continued, a kernel for Sundog, a kernel for Steve, until the bowl was empty.

Again the sharing, not the food, was the object. We felt sure of this for two reasons, first because Sundog didn't especially like popcorn and under normal conditions—if for instance we put popcorn in his bowl—he wouldn't eat it, and second, because Sundog in his early years was far too well mannered ever to ask at mealtimes for snacks from the table. While we ate, he would lie quietly in the corner and never look at us, let alone nudge us or whine or beg for a handout. At dog dinnertime, however, when he heard the can opener or the kibbles in the bowl, he would rush to the kitchen dog-style, his nails scrabbling on the slippery floor. For popcorn, however, he did nothing of the kind. Instead, when he heard the popcorn popper, he would enter the kitchen with a slow, dignified walk, possibly because he had noticed that human beings don't run into the kitchen for their food, or possibly because the popcorn as such was not important to him. He wasn't coming to the kitchen to eat, he was coming to share.

That sharing was his aim was confirmed beyond a doubt when we, the human beings, made a great mistake and spoiled everything. One night Steve was eating popcorn while reading a book. Sundog, with his usual quiet dignity, came into the kitchen to share as always, but that night Steve wanted to read without interruption, so when Sundog climbed into the chair and looked at Steve, Steve said, "No," and put several handfuls of popcorn on the floor.

Sundog stared, dismayed. Steve said, "No," again, more firmly. Sundog sat uncertainly for just a moment, then, apparently ashamed, he lowered his ears and his head and came cringing down from the chair to walk straight past the popcorn. Without a glance at us he left the room.

Horrified that we had seriously hurt his feelings, we followed him, bringing the bowl of popcorn and calling him to return. We found him in Steve's office, curled miserably on the couch. We offered him popcorn but he turned his face away from the bowl. We saw that we had humiliated him. We saw that we had shown him, bitterly, that he really wasn't one of us after all, that we were people, that we controlled everything, that he was just a dog, that he had presumed beyond his station in trying to be like one of us. He wouldn't look at us, he wouldn't come to us, he wouldn't touch the popcorn, he never returned to the table to share no matter how we tried to tempt him, *and for the rest of his life he never ate popcorn again.*

Even so, Sundog's ordinary behavior showed how deeply he had come to feel that he, Steve, and I were a group—the mixed, or dog-human group, as opposed to the other group, the all-dog trio of Suessi, Inookshook, and Fatima. Sundog seemed hardly to notice when two of the other dogs died, first the rickety Suessi, then his mentally absent former consort, Inookshook. In contrast to Sundog, the surviving elderly dog, Fatima, mourned for both of them.

I thought that Fatima would at last form a group of sorts with Sundog, but that didn't happen, largely because Sundog by then had lost interest in her. And anyway, instead of joining

Sundog when her family was gone, Fatima joined me. All she wanted was to sleep near me, by night at my bedside and by day on a couch in my office, coming awake only before meals. As her diabetes worsened she grew increasingly feeble despite daily insulin injections, and one afternoon she simply disappeared.

She wasn't gone long before we missed her and began a frantic search. We combed the woods and called the police and our neighbors and put advertisements in the newspaper and on the radio. I happened to know a state policeman who worked with a bloodhound, and I beseeched this man to bring the dog to find Fatima. He told me he would like to but he couldn't—the bloodhound was trained to follow human trails and nothing else.

At that point it came to me that Sundog could find Fatima. Probably he even knew where she was. And he knew about searching. Just a few weeks earlier, Sundog and I had been standing on the lawn waiting for Steve to go walking with us. But Steve didn't come. I went back inside the house to call him. No answer. Sundog was anxious to start the walk, so I said to him, "Go get Steve." But instead of going into the house, Sundog ran around behind it, where I heard him offer a few separate, forceful barks—*Wah!* (pause) *Wah!*—as he did when he wanted a dog biscuit or needed someone to open a door. There was no mistaking his meaning—if he had seen something coming and was giving an alarm, his barks would have been faster and on a different note, *Wahwahwahwahwah.* Sure enough, Steve had been behind the house, not in it, and of course came obligingly when he realized that Sundog was summoning him. The experience told us that Sundog not only understood what I wanted, but had kept in his mind a mental map of Steve's whereabouts without seeming to do so, and thus knew where Steve was at the time.

So I called Fatima repeatedly in Sundog's presence to show him that it was she I wanted. I also held her bedding against his nose to remind him of her scent. I then set off for the woods and told him to find her. He didn't seem to get the point, however, and instead of leading me anywhere, he simply treated my transects of the woods as if I were taking an ordinary walk, and bounded around me amiably. For three days we did this, then gave up. We never found Fatima.

It wasn't clear to me why Sundog let me down. Perhaps he didn't understand my request although he'd understood similar requests in the past. Perhaps he didn't think of Fatima as a group member (after all, she wasn't). So perhaps he couldn't imagine why I'd want to find her. Or perhaps he knew that she was dead and therefore held no importance.

I'd known other dogs who took little or no interest in the corpses of their deceased friends—after acknowledging with a sniff the presence and surely also the identity of the corpse, they ignored it. This is not to say that they didn't miss the dead—years earlier, on the occasion of the death of Suessi's brother, my dogs howled all night, sending their voices as far as they could as if, wherever he had gone, their calls would reach him. But it was he they wanted, not his corpse. To all the adult dogs I've had occasion to observe, a dead body, however dear its former occupant, is not the same as the living, breathing creature they once knew. (Puppies are different, and will cling to a dead mother's body as if to the mother herself, suggesting that as dogs get older they realize the significance of death.) So if Sundog realized that Fatima was dead, he might not have thought of her body. Fatima herself was elsewhere. If a corpse lay in the woods, covered with leaves, what was that to us or to him?

5

DESPITE his affection for us, from the very start of our relationship Sundog often seemed lonely. Because he was young and we spent our days in our offices, we weren't ideal companions for him, and he surely would have welcomed more excitement in his life than we provided. He took enormous pleasure in accompanying me to a pond at night where I watched a colony of beavers—on moonlit nights in the fall I would dress in black clothes including gloves and a ski mask to make myself invisible and I'd sit motionless under a tree while Sundog circled the pond and barked at the beavers. Unaware of me, they focused on him, showing how calmly and efficiently beavers react to an intrusive predator.

Sundog relished these forays. Every fiber of his being was alive and awake—he could have circled the beaver pond forever. Regrettably, though, we couldn't do this very often—the beavers after all needed to be left alone to work on their dam and establish their food supply for the winter—so unless the moon was full or nearly full Sundog and I stayed at home. Since Sundog at the time had little else to do for excitement, he often would ask to go to the pond, looking at me brightly with his ears high, ready to dash for the door at the slightest sign from me that we'd be going, but I'd refuse him, and he'd sigh with disappointment and lie on his mat to lose himself in dreams.

Mostly, he provided his own entertainment. One night during his first winter with us, I noticed him on a snowy hillside paying great attention to what appeared to be the burrow of some little creature. At the spot he sniffed and pawed, sniffed and pawed, then suddenly, as if a little animal had come out of the burrow, he rushed off with his nose to the earth to run in a great circle that took him back to his starting point. It seemed that the little animal, whatever it was, had returned to the burrow. But then, surprisingly, Sundog repeated his actions. Again the little animal seemingly popped out of the burrow to lead Sundog around in a circle. After absently watching this happen six or seven times it struck me that something out of the ordinary might be going on. Not many small animals would leave a safe burrow just to amuse an excited young dog, and fewer still would do so repeatedly. So I went to look, and sure enough, I found no burrow of any kind and no tracks in the snow but Sundog's. He'd been amusing himself with a fantasy.

Sundog also masturbated. We'd notice him standing still, moving his hips in gentle thrusts as he developed an erection. Presently he would ejaculate slowly, as dogs will. While he did this, though, his face would wear an expression of concentration. He'd furrow his brow and fix his eyes on the middle distance as if he were watching something. Was he having a sexual fantasy? And if so, whom did he envision? Was she a real dog he might once have known? Or was she a dream dog whom his mind's eye conjured, bracing her legs, holding her tail aside, and smiling up at him over her shoulder?

As of now, only a dog would know the answer. Yet for dogs as for people, masturbation is a second-best activity. We should not have been surprised, therefore, at Sundog's behavior when at last he met a dog named Bean and the masturbation stopped.

Bean was an overweight spayed female beagle-cross about six years old, who, at least to my eyes, resembled our lost Fatima. She belonged to a young carpenter, and she accompanied him one summer afternoon when he arrived in his pickup to start work on an addition to our house that we were building for my mother. Bean and Sundog greeted each other cordially and followed us indoors.

Bean was a bit shy at first—after all, she didn't know our house, but soon she was exploring it, followed by Sundog, who eventually went outside via the dog-door. Bean didn't know how to use a dog-door. She tried to follow Sundog, but stood by the door uncertainly when it shut behind him. Sundog must have wondered why she hadn't followed, and came right back, moving quickly and lightly, and watching her with an alert face and bright eyes.

Obviously, he liked her. He placed himself directly in front of her and bounced gently, inviting her to play. She, in turn, seemed pleasantly deferential to him. She approached him with her elbows bent and her head low, then looked up at him with her ears folded charmingly and a little half-smile on her lips. Soon the two dogs were running everywhere together, chasing each other, bouncing, bounding, completely absorbed with each other and their games.

At the end of the day, Bean's owner was ready to leave and called her to the pickup. She was a good dog and ran right over, but she kept glancing back at Sundog, who perceived what was happening and sat on the lawn, alert to developments. A faint expression of puzzlement came over his face when Bean slowly climbed into the pickup. As the truck went down the driveway, Bean put her head out the window to look back. At the sight of her, Sundog jumped to his feet, but then, when the truck disappeared around the corner, he slowly sat down again beside the driveway, his rump lightly resting on the grass, his body tense, his ears high, as if he thought that Bean might soon return.

Because most of us pay much less attention to our dogs than we think we do, the feelings that dogs have for other dogs often escape us. Steve and I may have noted how Bean and Sundog played together, but their play seemed normal enough and we made little of it. It was Bean's owner who first noticed that in the mornings Bean would uncharacteristically race for his pickup, and as he drove, would seem anxious. Would she see Sundog that day, or would she not? When the truck made the proper turn toward our house, her anxiety would turn to excitement. Each time the truck turned in the appropriate direction, her joy increased, and when at last the truck turned into our driveway, she'd leap around inside the cab, trying to get out.

If, however, Bean's owner went elsewhere, she'd realize that the truck was taking the wrong turns, and would be crestfallen. At the first wrong turn, she'd cry from anxiety, and after the next wrong turn or two, as soon as she understood that she would not be seeing Sundog that day, she'd sigh heavily and lie down on the seat, depressed.

When Bean's owner told us of her behavior, I began to take more notice of Sundog, who, the next morning, didn't come in for his breakfast but instead sat beside the driveway, his eyes bright and his head high as he looked eagerly toward the road. A few cars went by, but he took little notice of them. Then at last he heard a vehicle that brought him to his feet. When it came up the driveway, he ran forward to meet it. Yes! It was Bean's.

Bean was standing on the front seat leaning out the window like Juliet leaning over the railing of her balcony. The moment the pickup crested the hill, before it could come to a stop, she jumped out, an eager and vigorous Juliet clawing the paint in her hurry to join her lover. The two delighted dogs rushed to each other, kissing and inviting each other to play, then bounded away into the fields to spend the morning together.

Because I wasn't observing dogs carefully at that time, but was working on an unrelated project while watching the construction of my mother's apartment, and also because I didn't then realize that the life of a dog in an ordinary household can be fully as complex and instructive to observe as the life of a wild animal or a dog in an all-dog pack, I failed to pay enough attention to what the two dogs did except to see that they were obviously enjoying each other. In the days that followed, I couldn't help but notice Sundog's soft, radiant expression when he heard Bean coming or his joy when he was with her. He assumed, wrongly, that we could hear the distant truck as soon as he did, and also that we'd realize the significance of the sound, so he'd give us a quick, delighted glance and run outside to stand waiting in the driveway. He was wrong about our hearing the sound, and even more wrong that we could recognize the motor. But he was absolutely right that we realized the importance of Bean.

Who wouldn't? Sundog's happiness was so great that we found ourselves rejoicing with him. True, we lost his company to a certain extent—his doggishness had returned to him, so that he gave up his human mannerisms, and he didn't hobnob with people, but instead waited by the driveway for the arrival of Bean so that he and she could be dogs together. So perhaps his human attributes were, after all, a testament to an enchantment. Perhaps he'd been a real dog all along, a dog under an enchantment that made him seem like a person. Clearly, the kisses he received from the soft pink tongue of his beloved Bean had returned him to his true and best state.

Sadly, though, this joyous experience was brief. As the work on my mother's apartment drew to an end, Bean came less often. On mornings when her owner went to other jobs, Sundog would wait for hours by the driveway. At the start of his vigil, he would pay no attention to passing cars that he knew were not hers, but the longer he waited, the more willing he seemed to entertain the notion that an oncoming car might be hers after all. As his hope overrode his knowledge, his excitement would mount. Perhaps he'd been wrong about the sound of Bean's car. Perhaps some other car was her car, or perhaps she was coming in a different car. If a long time passed without her coming, he'd seem to be telling himself hopeful stories. In the afternoon he'd give up, go slowly into the house to his dogbed and throw himself down.

When the work on my mother's apartment was done, all the builders moved on to other projects, including Bean's owner. For more than a week, Sundog continued his vigil on the lawn, but now, no trucks climbed the driveway. Gradually he seemed to understand that Bean wasn't coming back. His vigils became shorter, and eventually stopped as he entered a depression that seemed to age him. His walk changed from a youthful trot to an old dog's shamble and his appetite faded. At this time he somehow broke one of his teeth—we didn't know how—and he stopped eating altogether. He slept a lot too.

We called Bean's owner to discuss the problem. He agreed that Bean and Sundog had formed a deep attachment. He said that Bean too was pining, and sometimes ran away. He believed that she went out looking for Sundog. He told us that he no longer dared to drive past our house but would take another route lest Bean figure out how to get to Sundog and would make a nuisance of herself by visiting us. We told him we'd be delighted to have a visit from Bean, but, understandably, her owner didn't want her roaming.

We offered to take Bean in the daytime. We'd even go to fetch her, we said. But her owner liked her company and also,

I'm sure, didn't want to impose on us. He declined our invitation. We offered to buy Bean. Again her owner declined. He loved her too much to part with her. This of course was very natural. Who among us would give up a beloved companion merely to please someone else's dog? We offered to take Bean if her owner ever needed to find another home for her—if, say, he were to run up against a new landlord who wouldn't accept pets, or a new wife who didn't like dogs. He said he'd remember our offer, but doubted that the day would ever come.

I wasn't surprised. We probably would not have parted with Sundog, even for his happiness. But we were never put to the test. Bean's owner couldn't take another dog. So Sundog and Bean suffered quietly from their separation, each of them alone, each of them depressed. From time to time my travels took me by the house where Bean's owner lived, and if she was outdoors when I passed, she'd come bounding to the road, her face lit with an eager smile. I'd slow down, and her eyes would question me. She obviously knew my car, and surely she wondered if at last she would see Sundog.

She never did. Few dogs get to develop relationships with their loved ones. Dogs are slaves, born to do what we want, not what they want. Even on the rare occasions when we see with any clarity what our animals desire, our wishes have high priority, theirs have low priority, and we convince ourselves, however wrongly, that we know what's best for them.

6

IF WE couldn't have Bean, we decided to get another dog, someone to take Bean's place in Sundog's affections. So for a number of reasons, all of them in serious conflict with my own negative views of pedigreed dogs, we agreed that we would get a Belgian sheepdog. We began a search that at first yielded no dogs but revealed an astonishing number of bizarre beliefs about dogs and pedigrees, including a statement from one breeder that unless we could prove that we would keep the dog confined for life in a wire run, no breeder would sell a dog to us. To such people a purebred dog is inanimate, a collectible, an item to be kept like, say, a valuable teapot. A mongrel dog is dirt or vermin, and a cross-bred dog isn't much better. One woman who gushed over the AKC even disap-

proved of wolves. Since the breeding of wolves is uncontrolled by human beings, wolves aren't purebred, she said, and they don't have registration papers. I wouldn't have thought of wolves as mongrels living in the woods, but evidently she did.

In time we found a breeder in another state who had five Belgians, including a well-known AKC champion. The champion's adolescent daughter was for sale. We drove to the breeder's home to see her.

The members of the breeder's family had jobs which kept them out all day, during which time the dogs were kept in crates stacked one above the other inside a shed that joined the house to the barn. Early in the morning and again in the evening the family took the dogs out, one at a time, and put them into a triangular pen not three feet wide, paved with a cracked concrete floor and enclosed by a chain-link fence. There the dogs relieved themselves.

After their morning visit to this dog-toilet, the dogs were returned to their crates to spend the day. In the evening the parent dogs were brought into the house where they slept with the human family, but the other dogs—those who were not house pets but were kept for show or for sale—stayed in their crates in the shed. Among them was the dog we were considering.

When we arrived to see this dog, the breeder showed us to chairs in her living room and went to fetch the young female for our inspection. The breeder soon returned with our prospect on a short leash. The dog briskly stepped along on pigeon-toes beside and behind the breeder, her tail tight between her legs and her head carried low and turned toward the breeder. However, the dog's eyes anxiously searched the room. She seemed nervous and uncertain, as if trying to please, wanting to do what was expected of her, but not sure what that might be. My first impression of the dog was that she knew only that what was happening was unfamiliar.

The breeder handed me the leash, and the dog immediately plastered herself against me, raising her chin and pressing her throat and the underside of her jaw tight against my leg, meanwhile staring into my eyes, the very picture of a witless victim. “She likes you,” said the breeder.

I didn’t agree. The dog had given herself to me instantly and completely without as much as an exploratory sniff of my sandals. She didn’t “like” me. She was helpless and she knew it, and she had no choice but to immediately appeal for tolerance. She’d been forced to sit beside me, so I became the target of her appeal. Interestingly enough, however, this dog kept her eyes on mine and she never even looked at her owners.

I asked how old the dog had been when she started her sojourn in the crate. She was crated even before she was weaned, answered the breeder. I asked if she had been confined by herself, or with her litter mates. All dogs were crated individually, said the breeder, as if this were a positive statement. And anyway, explained the breeder proudly, the litter mates of our potential purchase had been sold as young puppies because they were very desirable dogs. The reason the dog under consideration was still available at the age of seven months was because she had white fur on some of her toes, something which perhaps would not exactly disqualify her from the show ring, but which would never be to her advantage. Thus the breeder was asking only $200 for her. The “better” pups had cost much more. The breeder then announced that if we bought this dog we would need to produce a certificate of spaying from a veterinarian before we could register her with the AKC and receive her registration papers.

Excuse me? What had the AKC to do with us? And what would we want with papers? We would need them only if we wanted to show the dog or to breed her with another pedigreed dog to produce pups that could themselves be registered—something that the breeder obviously didn’t want us to do. But neither did we. We had already told the breeder that we had no more interest in breeding dogs than we had in showing them, and with or without inducement we would want to spay the dog, as we soon did. We did not, however, reassure the breeder by promising to send her the certificate (nor did we bother to send it when the time came). Why are dog owners so smitten by pedigree that they want registration papers for no reason? We pondered this question briefly, but lacked the curiosity to ask.

The dog in question continued to press herself against me, gazing at my face with anxious eyes. Her shape was that of a classical collie, and her beautiful black hair was shiny, so she was certainly good-looking. But her clinging display of cloying, helpless need was disagreeable. If she had been a person, I would have wanted to pry myself loose and leave her presence, never to return. But she wasn't a person. She was a sheepdog who lacked a meaningful life, who had nothing to do and no companions either. She was certainly not the dog of my dreams, but I knew that with a little work, the mental state of almost any dog can be improved, even a dog as needy as this one. Anyway, I couldn't leave her there. So we bought her.

Then we left. As we went out the door, the dog never looked back at the people who had raised her. On the other hand, she didn't seem happy to be with us either. Obviously she knew so little of life that she reacted to each new event with a disturbing mixture of anxiety and compliance, as if she knew no other way. She let me put her in the car because she didn't know how to stop me, but she seemed flustered and fearful, so I sat with her in the back seat intending to reassure her.

As soon as the car began to roll she looked anxiously down at the floor. It was moving and so was she. Why? She tried to jump up, to escape the unnerving motion, but had nowhere to go, so she flopped down on the seat again, twitching. All the way home she was uncomfortable and edgy, tensing each time the motion changed, drooling with carsickness and misery. And since I, not the breeder, had correctly guessed her motives for her earlier, clinging behavior, she wasn't at all soothed by my presence. Seeing me as the cause of her discomfort, she shrank away from my hands.

Worse yet was Sundog's reaction when, late that night, the long drive ended and the carsick, drool-dampened sheepdog cringed onto the dark lawn. Sundog of course investigated her vulva and anus while she politely stood still with her tail like the handle of a teapot, the midsection raised so that Sundog could inspect underneath, but the lower part pressed tight against her leg. Clearly she was guardedly pleased to meet another dog at the end of her dreadful journey. But Sundog's investigation was brief, and when he'd finished, instead of inviting her to play or showing in any way the delight we expected from him—the joy at finding an attractive female dog to take Bean's place in his affections—he turned away, and marked a bush nearby. He wanted to show the interloper that he, Sundog, owned the property. Then he left, showing no further interest in the new dog, nor caring why she'd come.

Poor, victimized sheepdog—AKC beauty though she may have been. What had Sundog seen in her that turned him against her immediately? Some so-called experts might suggest that, not being in estrus, she wasn't interesting to males. But they would be wrong. Dogs want much more in their mates than the mere whiff of estrus. Sundog had desperately wanted Bean, and Bean was spayed. Could Sundog have learned in the first few seconds that for all the new dog's beauty she was ignorant and emotionally damaged, nothing at all like his competent Bean? Was it simply that she wasn't Bean?

And what about the new dog? Upon leaving the car she'd seen a strong white dog radiating presence and authority, and immediately she wanted to be his companion, to accept him as her leader, only to have him walk away, indifferent to her feelings, ignoring her situation, and casually lift his leg to post a Keep Out sign. If these two dogs had been people, he would have slammed a door in her face.

So it wasn't hard to imagine how she felt. In my first year of high school, I'd felt the same at my first dance. I'd worn flat black shoes, semi-opaque stockings, and a white cotton dress with puffed sleeves and a long, loose skirt with a peplum—an extra flounce that hung over my rump. My elders had chosen my clothing. Most of the other girls wore upswept hair, eye shadow, nail polish, earrings, high heels, and strapless satin gowns with tight waists and full skirts of gauzy nylon net. I'm sure I looked the way I felt, awkward and unhappy. Predictably, the boys ignored me just as Sundog ignored the new dog.

So perhaps Sundog's rejection was understandable if not kind. The new dog was nothing like Bean. Still, I was disappointed. Even so, I could certainly excuse him by imagining myself in his situation. How would I feel if some powerful creature took away my husband, the loving, familiar man of my choice, only to bring into my house a man I'd never seen before—some incompetent stranger from the street? *Here! Take this one for your husband!* I'd be horrified, of course, and my first act would be to show the stranger that the house was mine, not his, and he'd better understand that he should keep his distance, that I wanted nothing to do with him.

This proved to be a fair analogy. As any of us might have done under similar circumstances, Sundog seemed to hope that the intruder would leave, or so we thought the following morning, when he showed her his teeth.

She, on the other hand, was touchingly attentive to him. When mealtime came, she ignored the bowl of food we put down for her and assumed a charming, pigeon-toed stance near Sundog, who was busily eating from his own bowl. There she cocked her head, put her ears forward, and fixed her eyes on him. Had she never seen another dog eating? Perhaps not. Her crate had been against a wall.

Sundog didn't care one bit for her inspection. Turning to face her, he flattened his ears and threatened her. Dismayed, she took a step back. Sundog then ignored her. When he finished his food, which he did very quickly as always, he left the house by the dog-door—invariably his second act of the morning. The new black sheepdog would have followed, but the flap of the dog-door slammed shut in her face.

Thus the new dog met her first challenge. Startled, she waited for the door to open for her too. When it didn't, she waited a moment longer, then turned back, dejected. She was not going to be given a chance to go outdoors, she saw. People controlled the opening of doors, in her experience. She didn't understand that the little, low door was meant to be opened by dogs.

Later that day I named her Misty, as if to counterbalance Sundog. In contrast to him with his white fur, his strong, clear mind, and his well-defined persona, she was a dark, soft cloud of a dog who had been born to live imprisoned and alone, so that she and her world seemed obscure.

I soon began to realize the extent of Misty's limitations, something that Sundog may very well have noticed on his own the night before. Very soon I learned that Misty couldn't walk through grass. When she tried, she actually stumbled, then stood still, embarrassed and dismayed, because she didn't realize that if her feet were *in* the substrate, not *on* it, she should lift them with each step. But how could she know? She had never walked on grass before. Her only visits out of doors had been to the little, paved latrine.

She also knew nothing of stairs, and would not attempt to climb them. The night we brought her home we had to carry her upstairs so she could sleep with the rest of us in the bedroom, but in the morning she wouldn't follow us down.

She couldn't. Instead, radiating helplessness and fairly weeping with agitation, she begged us with her entire being—her voice, her eyes, her posture—for something she desperately wanted, but what? To be carried? To make the stairs vanish? To change our plans and spend the day with her on the second floor? When we didn't immediately help her, but waited to see if she'd figure out the stair-climbing process for herself, she leaped clumsily toward us from the top stair as a clown might leap outward from a diving board. She fell into the middle of the staircase and slid the rest of the way. After that, until she learned, I carried her.

The dog-door mystified her utterly. She saw how it opened readily for Sundog, letting him pass through freely whenever he chose. He had merely given it a little push, of course. But when she herself approached it, it mysteriously stayed shut. Easily discouraged, as if a negative result was the result she'd expected, she would not even approach the dog-door after her second failure. We saw that Misty didn't have the slightest sense of self-worth or self-empowerment, and we would need to start her education with the basics.

We began with the dog-door. A rectangular opening cut into the shed door, the dog-door was covered with a heavy plastic flap that swung both ways. When the flap was not in motion, magnets held it tight within its frame, keeping out mosquitoes and the weather. I meant to teach Misty to push herself through the opening, no matter how forbidding the closed flap might appear. The closed flap, then, became her first challenge.

Leaving Misty in the shed, I went outside and shut the door behind me. Misty wanted to follow—I could hear her despondently nosing at the plastic flap, which I then held aside. The opening framed her dark face and searching eyes. With encouraging words, I called her. She stepped through, and was outside. We repeated the lesson with the flap open a little less each time so that Misty had to push a little more. By the second day she was all but opening the dog-door by herself, and by the third day she had mastered it, although I still needed to call to her from the opposite side of the door before she'd try.

When we weren't working with the dog-door, we worked on going up and down stairs, starting with up. Although trembling and panting, Misty did reasonably well, getting her feet more or less on the treads until she was about halfway to the top, at which point she collapsed from lack of confidence. Reclining on the steps she had already conquered, half on her back with her head low and a forefoot raised in a submissive effort to appease me, to prevent my forcing her to continue with the lesson, she slowly slid downward until stopped by the wall. When she could slide no farther, she seemed relieved.

After she had a rest and a snack, we tried again, and again that evening. At last she seemed to understand the principle—you bravely step upward on alternate forefeet and just keep going. Getting the hind feet right was the easy part—the action of the hind feet took no thought or willpower but just came naturally. After an entire day of practice Misty climbed the front stairs unaided to accompany us to the bedroom.

In the morning, though, she cried for help to follow us down. Interestingly, the simplest principle that applied to going up seemed irrelevant under these new circumstances. I tried to show her how to move her feet, but she panicked as if such a thing had never been asked of her before—as, indeed, it had not. But we persisted, she and I, and eventually she learned how to go down the front staircase as if descent were a whole new skill.

Then came the back stairs. To my great surprise, Misty could not apply the principles of the front stairs to these new stairs. Instead, she stood at the bottom, forlorn and dejected, once again believing that the challenge she faced was too great. So, starting fresh, as if I'd never taught her anything, I then taught her how to use the back stairs. First she learned how to go up, then she learned how to come down.

I also taught her how to use a very short flight of stairs to my office, which was in an outbuilding at a distance from the house. Here again, she felt overchallenged, although perhaps she learned the up part and then the down part a bit faster. When finally she'd mastered these stairs too, we thought she was at last equipped with a little independence. We soon saw how wrong we'd been—yes, she could leave the house via one dog-door just as we had taught her, but, at least for a while, she couldn't get back in even though the action was the same in both directions, nor could she use the dog-door to my office either to go out or to come in. As with the stairs, Misty had to learn how to master each direction of each dog-door as if it were an entirely new skill.

All in all she learned the following skills separately and at different times: how to enter the house via the dog-door, how to leave via the dog-door, how to enter my office via the dog-door, how to leave via the dog-door, how to go up the front stairs, how to go up the back stairs, how to go down the front stairs, how to go down the back stairs, how to enter my office via its own small set of steps, and how to leave my office via these steps—ten new skills altogether. This list does not include walking in grass without tripping because Misty figured that out for herself. The list also doesn't include negotiating a flight of extra-steep stairs to the attic above my office—the original stairs put in when the building was a farmhouse—because Misty never learned to climb these stairs and never visited the attic. The very sight of a dog or even a person using these extra-steep, frightening stairs made her uneasy.

But then, no staircase pleased her. Years later she would still toil up or down any set of stairs with an expression of deep concentration on her face, her head and tail low, her shoulders hunched, her ears folded, and her feet working busily as if the unpleasant task required her full attention. I'd known dogs who would fearlessly climb anything, even trees or ladders. Misty wasn't one of them.

What was Misty's problem? How was it that a healthy, full-grown dog could view the act of walking through grass or opening a dog-door or climbing a flight of stairs as an insurmountable problem? And how was it that she couldn't transfer what she'd learned from one task to a similar task? She had no physical disability—in fact, as soon as she learned to pick up her feet when traveling through grass, she could run as fast or faster than any dog I'd ever known, including my champion dog-sled racers. Her disability was purely psychological.

She wasn't stupid, although her inability certainly made her seem stupid. And she liked to learn. In fact, she learned some skills very easily, thus proving herself highly trainable. Ordinarily, I don't give my dogs much training because I want them to do their own thinking, to do what they want rather than wait to see what I want. If I train them, they learn from me. If I don't train them, I learn from them. So most of what they needed to know, they learned from other dogs, and they usually performed to perfection sooner or later. Such is the power of group membership, especially if one member doesn't tyrannize the others. I do however teach my dogs five things: (1) the meaning of *no,* (2) to come when called, (3) to urinate and defecate outdoors only, (4) not to take our food, and (5) when we're taking a walk, never to chase a car but rather to come to my side and sit while the car goes by.

Misty did very well with all that. Having been crated, she could wait more than twelve hours without relieving herself, not that we would ever have asked that of her. And after just one lesson at roadside sitting she performed brilliantly. Tempted by her trainability, I also taught her to heel, which she then would do faithfully whenever asked. She even liked to heel, and soon, on our evening walks, she would keep close beside me, her head high so that my fingers could brush the tips of her silky, upright ears.

But her capacity for learning certain motor skills remained compromised. And the cause of this seemed clear enough—her failures were due to her childhood confinement. Like young people, young dogs have a latency period during which they learn how to relate to experience and also how to cope physically with their environment, and if they miss the opportunity, their learning is blighted. Alone in a crate, she never had a chance to gain the kind of education that other puppies acquire almost unconsciously.

7

SOMETIMES I thought I could see what the first few months of life must have been like for Misty. At times she would slowly lick her wrists or ankles, a practice which, if she kept it up, made patches of raw skin. Perhaps this behavior originated when, without toys or companions, without parents or people, this young, inexperienced creature had nothing but her own body for consolation. Perhaps the feeling of a tongue against her skin reminded her of the comfort she would have known only during her first few weeks of life, before she was crated—the comfort of her mother.

After she had been with us a few months, however, the difference between her former life and her new life began to dawn on her. One morning she was standing beside me at the kitchen

door, humbly waiting for me to let her out, when an idea struck her. She gave me a quick, bright glance, then dashed into the shed and leaped through the dog-door by herself. From that moment on, she passionately preferred the dog-doors to the regular doors, which—unlike Sundog, say—she scorned. After that, for many years, every time the dogs and the people would prepare to leave the house together, Misty would see what we were doing and would rush to the shed, where she would sail triumphantly through the dog-door. She would then run around the house and, radiating delight, would meet us as we came out the kitchen door. She even became stubborn about this—she would insist on using the dog-door, and would never use the regular door even if we tried to make her. No longer a helpless, needy victim awaiting the pleasure of human beings, she seemed to exult in making her own decisions and having control of her life. The dog-door became her triumph. It empowered her and in many different ways it set her free.

At about this time I learned of an experiment performed on a white-footed deer mouse confined in a box where he could control the lighting. The purpose of the experiment was to determine whether such mice prefer darkness, twilight, or full daylight. What the experiment proved, however, was that this mouse, at least, didn't want someone else trying to run his life. If the experimenter turned the lights up, the mouse turned them down. If the experimenter turned the lights down, the mouse turned them up. The mouse wanted control, and he took control. In this, the experiment reminded me of Misty.

Another event was also of interest. One day, I was working in my office when I heard a faint scrabbling sound behind me. There lay Misty flat on her side on the bare wooden floor, staring into space and absently scraping at the floor with her teeth. Evidently this was what she had done when locked in her crate without toys or companions, lonely and bored and probably teething. I felt a sudden rush of sadness for her. Into my mind came the following thought: *Sweet Misty,* I imagined myself saying to her, *you're free now. You don't have to gnaw at the floor ever again. If you like, you can go outside and play.*

Then a surprising thing happened. No sooner had I completed this thought than Misty raised her head and looked up at me. Then she suddenly leaped to her feet and ran out the dog-door to a nearby maple tree where she found a stick that she began to shake and maul. It was as if the idea had come to her the moment it had come to me. It was as if Misty were a person who had heard me speak the thought. Her response could not have been more immediate or appropriate. I was very impressed.

But I wasn't surprised. It has often seemed to me that dogs and people can sometimes read one another's thoughts accurately and in detail, just as dogs seem to read the thoughts of other dogs and people seem to read the thoughts of other people. What's more, over the years I've had similar experiences in which animals also played a role, experiences which some might say were of a psychic nature. These will be discussed at the end of the book.

Misty never gnawed at the floor again. And the moment she stopped gnawing the floor was the moment she saw how to use her freedom. The dog-door showed her how to leave the house, and the stick under the maple tree showed her that she was free to play once she got outside. What's more, being a dog, and having the wonderful, doggish ability to compensate for deprivation, she showed fewer signs of her early imprisonment than one might suppose.

However, not long after she learned about freedom, she began to chase cars. None of our other dogs had ever taken an interest in cars, so we were unprepared. One day we noticed that Misty had established a sort of lookout for herself, a mossy outcropping near the house from which she could watch the road. She'd lie there quietly enough in her soft cloud of fur, but she'd keep her head and ears up and her eyes open. When she heard a car coming she'd jump up and ready herself, holding herself back as if not to startle it. As it sped past she'd rush it, then tear along beside it, lunging at it as if trying to nip its legs and make it turn. Misty could run at speeds of twenty-five to thirty miles an hour, much faster than any of the rest of us, and almost as fast as a car could travel on that particular road. She'd be gone a long time before we'd see her again, smiling, almost dancing, tail waving, radiating joy and ready to chase the next car.

Of course, we couldn't let this happen. We tried to train her not to chase cars, but without success, and eventually were forced to make her wear an electronically activated shock collar that gave her a warning buzz, then a shock, if she approached a buried electric wire on her way to the road. Fortunately, though, the collar didn't curb her sense of freedom. It merely stopped her from chasing cars.

It may, however, have increased some of her anxieties. There she'd be, ready to chase a car, but knowing that a mysterious force would hurt her if she tried. As if in a crate, she was again at the mercy of something she knew nothing about. She then developed a few minor ways of coping, one of which was that she began to restrict herself. She wouldn't allow herself to relax just anywhere, for instance, but only in four places—a certain chair, a certain end of a certain sofa, a certain area of the kitchen floor under a window, and a certain corner of our bedroom. Also, when leaving the porch, she wouldn't step straight off the front like everyone else, but insisted on walking its full length and stepping off the side. And as she went along the porch, she always avoided the middle or the open side and unfailingly hugged the wall. She had her own special route around the front yard too, a route for which no practical purpose could be found, and she'd approach the house only by a certain walk, even if to do so took her far out of her way. As she performed these acts of illusory reassurance, she resembled a person knocking on wood or tossing salt over one shoulder, or walking carefully on a sidewalk to avoid the cracks.

Her ambient nervousness wasn't entirely unfounded. Control of one's life implies good management, and Misty didn't have enough experience or flexibility to be certain of achieving that. Occasionally something would happen that she couldn't cope with, and then her sense of helplessness and worthlessness would overcome her anew. One day in early spring of the first year Misty was with us, Steve, Sundog, Misty, and I all went for a walk together. Misty and Sundog rushed into the woods, as was their custom. But when we called them back, only Sundog appeared.

Early spring is a dangerous time for rural dogs. The ice is melting on the various bodies of water. Every year in these parts, dogs fall through and drown. Our old dogs, the wolfish Suessi and the dingo-dogs Fatima and Inookshook, never had a problem with rotten ice—like most wild animals, they seemed to know when ice was safe and when it wasn't. They never walked on black ice, for instance, and even seemed to know when snow-covered ice might be weakening. Sometimes they seemed to be testing the air and also listening when near it. Perhaps they were searching for an odor. A stronger pond-smell, maybe? Or perhaps rotten ice produces, or fails to produce, a characteristic sound, perhaps when splashed by the water under it. Sundog also managed himself safely when crossing ice-covered ponds.

But Misty was different. We became alarmed when a few minutes passed and we saw no sign of her, so we split up to check the nearby ponds. As we were about to leave the area to check more distant ponds, we heard a moan—a low, muted, strangling call as if forced from a rigid throat.

We followed the sound and found Misty not fifty feet away, standing chest deep in a beaver pond that was right beside the road and so shallow that we hadn't thought to search it. She seemed unable to move, so we waded in to lift her, and found that she was almost in shock. Her body was stiff and her eyes didn't focus. We set her on her feet but she collapsed. She didn't seem to recognize us, or know that she was out of the water.

From her tracks in the snow we saw that she'd run across

the pond to a place where a brook fed it, where the ice is predictably unsafe and where our other dogs would never have gone. There she had broken through. The pond was so shallow that other dogs would have clambered out and gone on their way, but Misty hadn't tried to get out. Stunned by what had suddenly occurred, she had again become a helpless victim, numbly waiting for whatever might happen next. Fortunately she'd managed to moan.

We rushed her home to a warm bath, then to the vet for a checkup, not that she'd been in any real danger, as only her legs and belly were wet and she hadn't been long in the water. So she quickly recovered, and may even have learned from the experience, as she never had a similar accident, but never was the difference more clearly shown between Misty and dogs who know how to cope, dogs who learn how to handle themselves early in life, dogs with experience and confidence.

Not surprisingly, Misty did poorly on the towel-biscuit test. She liked dog biscuits, and when I showed her one, her ears went forward hopefully and her mouth watered. But when I put the biscuit on the floor and covered it with the towel, she seemed dismayed. I expected her at least to sniff around the edges of the towel, but no, she stood transfixed, glancing first at me, then at the towel, then at me again. Her attitude was reminiscent of her first approach to a staircase or a dog-door—she seemed to see an insurmountable obstacle against which she was powerless. *I don't know how to cope with this and no one will help me,* her manner said. When I saw she wasn't going to try to get the biscuit, I lifted the towel to let her have it, but even when it was in full view, she didn't take it, and continued to stare at it with awe. She seemed to associate the biscuit with her inability to manage. She'd been made to feel inadequate. She'd been humbled by the test. I picked up the biscuit and put it crosswise in her mouth, but still she stood transfixed, drooling, as if she didn't seem to know what to do next. I patted her and spoke to her in a light, carefree tone until her frozen, helpless feeling suddenly lifted. Carrying the biscuit, she quickly left the upsetting scene to eat it elsewhere.

Was she stupid? Not at all. She had learned to heel and other similar skills almost without effort. Later in life, when she became hard of hearing, she learned hand signals with no training at all, merely by once observing me make a circular motion of my hand when I was trying to get her attention. Taking this as a signal to come, she trotted right over, shy and happy, head low, ears folded, and tail modestly waving as if she were delighted to know what I wanted and could please me. After that, we simply used the circular motion as a signal to come. It was that easy. It was as if she understood her hearing problem, and had herself been searching for a solution. And if that isn't a sign of brains, I'd like to know what is.

So the towel-biscuit test proved nothing about Misty's intelligence. Who among us cannot remember being confounded by something, especially in our young years? I can still remember the numb fear I felt in fifth grade, when the Latin teacher entered the room. I knew he'd call on me—he always did. I knew I'd stammer and stare at him, eyes wide and mouth open, while all the other children sat in frozen silence, terrified for me but pulling for me in their hearts. Half a century later I remember one of the words the teacher asked me to translate—*aedificat.* I hadn't a clue. *"Edifice!"* he shouted. *"Edifice!* An imposing *edifice* stood on the corner of the block!" But I'd never heard of an edifice, wasn't sure what "imposing" meant, and couldn't answer. Besides, I thought he'd said "corner of the building." Perhaps I wasn't the smartest kid in the school, but in that instance at least, my ignorance wasn't the only problem. Thus Misty must have felt when she saw the towel go down over the biscuit.

Otherwise Misty lived her life among us, accepting the fact that Sundog didn't much like her, feeling content to be free of the crate and part of our group—a participant in our mixed-species relationship. By day she took control of her life as best she could, using the dog-door, running skillfully through grass, playing when she felt like it, and observing the little compulsive acts that made her feel secure. Now and then she'd visit me as I worked in my office, just to nudge me for a pat, just to be sure I was still with her.

I think that she preferred the nights, however. At night the members of her group were relaxed, going nowhere, doing nothing challenging or difficult. Also all of us were in the same room, close together—Misty on the floor at my side of the bed, Sundog at my husband's side, the cats in their familiar places on a chair, on a windowsill, in the laundry basket, on a closet shelf, on the foot of the bed. When night came, Misty was first up the stairs.

One night I heard Misty stand up and go quietly downstairs. The dog-door clicked, so I knew she'd gone out, probably to relieve herself. Soon she came back, her fur cold and wet. Although I didn't move or speak, she knew I was awake, as dogs always seem to, and she came quietly to the bedside and put her head on my shoulder. As she pressed her face against mine, I put my arm around her. A wave of affection passed between us. On the far side of the bed, Sundog felt it, and thumped his tail gently, as dogs sometimes will when affection is displayed among group members. And in his sleep, my husband also felt it, or perhaps heard Sundog's tail, and without waking he turned to put his arm around me. Thus some members of our group expressed their feelings for one another. But we were not the only creatures in the room who experienced the moment—from the foot of the bed, a cat also joined us. Without moving or turning or looking at us, he very briefly purred.

8

DOGS are very much like people in their social placements. In some human societies, as anthropologists know, the point at which a person enters a group determines his social position. Some societies hold, for instance, that a baby born into a family in June, say, outranks an adult who marries into that family in July, simply because the baby joined the family first.

Dogs are much the same. Many people who acquire a second dog observe that the first dog continues to outrank the new dog even if the new dog is bigger and stronger. Both dogs know who got there first, and both understand the first dog's seniority. There are many variations on this theme, and sometimes the new dog takes over, but not nearly as often as one

would expect if youth, size, and strength were the controlling qualities. To dogs, seniority means more than many of us realize. It's a wolf thing.

On a visit to Scotland, I once saw a group of captive wolves who were kept in a many-acre pen, a group from which one member had been excommunicated. He was thin and weak and his fur was matted—obviously the other wolves weren't letting him share their food. During the many hours that I watched these wolves, the outcast hid in some long grass, crouching low, pressing himself against the fence but keeping watch on the others so that he could stay as far away from them as possible. If the other wolves noticed him, they ran down and attacked him. Hence he was not only thin but also wounded. His keeper should have removed him. In the wild, he surely would have left the area and gone far away to fend for himself as best he could. Trapped in the pen, he was at the mercy of the others.

What explained his situation? I wanted to ask about him, but the keeper wasn't present. Still, there were several possible reasons for his plight. Perhaps he had never been a pack member, but had been shoved into the cage as a stranger on the erroneous assumption that the others would accept him. *He's a wolf, the others are wolves, no problem,* the theory goes—the same theory that thrust Misty and Sundog together.

Or perhaps the Scottish outcast had fallen in rank. Perhaps he had challenged the leader and was defeated, or perhaps he had occupied a middle position and was unseated by an ascending wolf. Such dislocation can be disastrous in wolf society, where often a wolf falling from its social position doesn't merely slip down a rung or two, since these rungs are already occupied, but slides all the way to the bottom if not out the door.

Perhaps this is due to the way wolves perceive one another. When wolves such as those of, say, *The Jungle Books* look at their pack mate Grey Brother, they may not see him merely as an individual, but as the personification of Place Two, the rank he occupies. Thus, if another, upwardly mobile wolf dislodges Grey Brother, Grey Brother escapes into the woods while Place Two continues to be personified. Who cares what happens to Grey Brother the individual? Place Two is filled and that's what counts. We experience much the same feelings for elected officials—we respect a person and accord him high rank because he is, say, the president, not necessarily because of his personal attributes. In animals, the attitude tends to stabilize group structure, which is as critical to wolves as it is to us. "The king is dead, long live the king" is a concept wolves could live by.

As for the starving, wounded wolf crouched by the fence in Scotland, grim as his situation was, it clearly showed what happens to wolves without status. His plight and that of others like him explains why the possibility of falling in rank seems especially terrible to many dogs.

It certainly did to Misty. At our house, she gratefully accepted the position of Dog Two. She already knew from personal experience what it was to be a dog with no rank at all, and was so pleased to have status and to belong among others that she treated those around her, including the cats, with deference and grace.

However, her apparent ease was deceptive. Foremost in her mind was her wish to keep her place, and the possibility of losing it preoccupied her. To her, any outsider was a threat, so she followed and harassed all dog, cat, or human newcomers, making every effort to get rid of the unwelcome guests lest one of them should join the group and challenge her position. She would even display her rank to people. Who knew which of them would join our group and dislodge her? Showing that she had more on her mind than merely alerting her group to the presence of a stranger (which was all that any of our other dogs had ever meant by barking at visitors), Misty would sometimes not bark at all when a visitor came, but would unpleasantly sneak around behind him to bite his leg. This dreadful habit, I felt sure, was meant to discourage him from staying, or even from approaching the house.

Misty's behavior with visiting dogs was even worse. Any visiting dog, whether young or old, male or female, she saw as a possible rival for her cherished Place Two, and she'd follow it about, barking viciously or standing over it to point her nose down at its back—a dominating gesture. Misty was big, and also, surprisingly enough, an excellent, fearless fighter, as some visiting dogs learned to their sorrow and as most visiting dogs perceived. Most of them cringed away from Misty.

Sundog, in contrast, was so confident of himself and his place in our group that he didn't seem to feel threatened in any way by visiting people or by most visiting dogs. He'd bark at first, then briefly investigate visitors of either species. If they were people, he would eventually court them politely, making sure that he knew them and they knew him. If they spent the night and didn't get up before dawn with the rest of us, he would lie in front of the guest room door until they did. On their subsequent visits, he wouldn't neglect to bark at them, but probably just on principle, and then, in a more or less perfunctory manner, he would go through his welcoming procedures again.

Mature male dogs seemed a bit more challenging. Sundog might bristle the hair on his hips and shoulders and pointedly mark a bush or two to remind the other dog that he, Sundog, had high status, but there his display ended. The visiting male would quickly take in Sundog's size and ambient authority, and would assume a deferential, pacifying manner, treating Sundog as a human guest might treat his host. Sundog's calmness gave the visiting dog calmness. In all the years we knew him, Sundog never had a fight.

Misty was different. She couldn't relax. Nothing showed her fears more clearly than her behavior with cats. She was touchingly deferential to the five cats already with us, especially to the high-ranking cats, Lilac, who was Female Cat One, and her son, Rajah, Male Cat One.* If one of them were lying in a doorway, for example, Misty might sit down to wait for the cat to get up before she'd pass through. In short, she treated the cats who were her seniors in the household in almost the same way that she treated Sundog. By dog law, she knew that they outranked her, and the cats agreed.

But now and then, a new cat would join us, and to this newcomer, Misty would react very differently. In her mind, the new cat surely wanted to unseat her, to take over her social position, and she wasn't about to let that happen. So, as the poor cat tried its best to adjust to our complicated household, Misty would harass it, following it around, barking at it vehemently and obnoxiously for hours on end, and pointing her nose down at its back with the same gesture that she'd use against a visiting dog. *I'm your social superior, and don't you forget it,* Misty's manner said.

This put the new cat in a difficult situation. To avoid stimulating anyone to chase it, the cat would want either to hide or to stand around calmly, pretending that everything was fine. But Misty made this impossible. If the cat was hidden, she'd find it. If it was trying to seem composed, she'd unnerve it. We'd call her off, but she'd sneak back to bother the cat some more, making sure it knew that she and only she would occupy her rung on the social ladder, and that the cat had better watch its step. A more secure dog wouldn't worry about the potential status of cat newcomers. Sundog, who didn't really like cats, completely ignored them. But then, Sundog knew his power. Misty had no notion of hers. Hence her constant worry. When I began to make arrangements for a third dog to join our household, I wondered what Misty would do.

*Lilac was mottled, and was named for her color. Rajah was named for a certain tiger in Illinois.

9

THE third dog lived in Boulder, Colorado, a member of our son's household. Among the nonhuman species in this household, the leadership was provided by a marvelous Australian shepherd whom our son had rescued from a pound, a dog named Rider. Rider's following included three other dogs and numerous cats, of whom Rollo, a large marmalade male cat, and Manas, an equally large Siamese cat, were paramount. Manas had denied herself nothing over the years and at last became so fat that she looked like a beanbag, and could assume any known position without toppling over. At the time, most of our cats were related to Manas—her children, grandchildren, nieces and nephews—as were the cats who lived with our daughter and her husband in Austin, Texas. Our son's family also owned a rabbit, who strangely enough had taken Rider as

his leader. Surprisingly, this rabbit sometimes acted like a dog.

Evidently, he had modeled himself on Rider. Wherever Rider went, there the rabbit went also. Few rabbits have been known to hunt squirrels, but this rabbit did, accompanying Rider on forays against the squirrels who ventured onto the property. Together, Rider and the rabbit would slowly stalk a squirrel, together they would rush it unsuccessfully, and together they would gaze up at it as it escaped into a tree.

Also in Boulder lived an Australian shepherd–chow cross, a beautiful female who often walked on a leash with her owner on the sidewalk that passed our son's house. Many times, Rider and this female had investigated each other through cracks in the fence, and when the female came into heat, Rider was ready. He jumped the fence and mated with her. Since the female already knew him, she was, I'm told, glad to hold her tail aside for that splendid dog.

Well, nothing could be done. If the Aussie-chow received mismating shots to abort the fetuses, the shots didn't work, and eventually she bore Rider a litter. Perhaps what happened next was normal considering the politically correct atmosphere of Boulder—the owner of the female held that when children are born, the father must assume his share of responsibility. One day the Aussie-chow's owner appeared on our son's doorstep with a female puppy. Our son's family took her in.

Rider, of course, was delighted. Perhaps he knew that the pup was his daughter, or perhaps not, but since he was Dog One in that household, and since the puppy welcomed his leadership, he immediately accepted her into the pack. Our granddaughter, Zoë, named her Pearl.

Pearl grew up to be strong and vigorous. Like her father, she could jump the fence, but unlike her father, she was never content to stay in the yard for any reason. Every morning she would liberate herself and run away. At the local humane society, her name became synonymous with pest and nuisance, and our son, who was quickly gaining a reputation with the local dog wardens, was at a loss as to what to do. Because of

Boulder's zoning ordinances he couldn't erect a higher fence around the yard, nor would his yard at the time accommodate a buried electric fence, so young Pearl faced a choice between impoundment at the humane society or permanent detention at home at the end of a chain. Meanwhile, her owners faced large fines and ongoing legal problems, or the guilt-inducing presence of a miserable, frustrated dog.

At this time I came to Boulder for a visit and met Pearl, who by then was almost one year old and almost grown. Blue-merle like her father, but built like her part-chow mother, she seemed as solid as a hydrant. On the rare occasions that she was not in motion, she seemed very firm on her small, white, widely planted feet, like a little table. When I first saw her she was shuffling through the snow, producing two parallel lines of tracks like a bear's rather than a single line like the tracks of other canids. Yet she was feminine and quick, a happy dog whose tail curled pleasantly upward. Her velvet ears were always raised, always listening, and intelligence shone from her beautiful black eyes.

Perhaps she saw how well I got along with Rider, her father. Perhaps she sensed how deeply I admired dogs in general and herself in particular. Perhaps she saw that I meant something to the animals and people of our son's household, and decided that I must therefore mean something to her too. Anyway, she chose me. She stuck to my side by day and insisted on sleeping on my bed at night. Her family loved her dearly, but at last decided to give her to me.

I'd already returned home by then, so they bought her a crate and a plane ticket to Boston. Our son's wife said something very nice about us when she took Pearl to the vet to get the necessary papers for the journey—she told the vet that Pearl was going to Dog Heaven. The vet was shocked. He liked Pearl—anyone who knew anything about dogs would like her—but having heard of her legal difficulties, he took the remark to mean that she had been brought to him for euthanasia. Not so. She was coming to me.

Pearl's journey must not have been enjoyable. I opened her

crate the moment I'd hauled it to the airport parking lot. Out came a rather flustered dog, who didn't seem to know me. Instead of offering me even a tentative greeting, she stood squarely planted on the pavement, looking slowly around in all directions as if trying to determine where she was. No helpless victim this, I realized—no dependent, needy creature who would turn to the nearest human for assistance. Pearl was going to get her bearings first, and only then might recognize me.

I showed her the car. Needless to say, she didn't want to get in. I gave her leash a little tug and spoke to her in a coaxing manner. She braced her legs and resisted. I had brought water and a bowl with me, and offered her a drink. She refused, and kept looking around. I lifted her into the car. She crouched in the far corner of the back seat, tense and glowering, and remained there all the way to New Hampshire.

Once there, however, her demeanor changed. Not at all intimidated by Sundog, she let him investigate her for a moment, keeping her head and ears low and waving the tip of her tail once or twice to show that she fully understood his rank and would of course defer to him—no problem. As if such encounters were routine to her, she spent a moment squatting on the lawn to relieve herself, then, without waiting while Sundog overmarked her stain, she trundled into the house as if she'd made up her mind to confront the new situation head-on, and wanted to get the introductions over with.

There stood Misty. Horrified by the arrival of another dog, she began to bark fiercely, then stopped as suddenly as she began when Pearl—her demeanor pleasant, her head and tail modestly low—went right past Misty to a bowl of water, despite Misty's warnings. *No problem,* Pearl's manner said. *I need a sip or two but don't worry, I respect you.*

Head high, ears up, Misty watched Pearl with amazement. Normally, a visiting dog would cringe away. Pearl's determined self-assurance must have puzzled her. Here was a dog who seemed to understand her low rank perfectly, but even so was confident enough to do as she pleased.

At this point, Sundog came in through the dog-door and calmly crossed the room as though Pearl were already accepted. That was enough for Misty. Dogs meeting for the first time usually establish their initial relationship quite quickly, as all dogs know. Misty wouldn't need to worry about keeping Place Two. In fact, Pearl behaved as if Place Three was all she wanted, as, indeed, it probably was. Evidently, in Pearl's experience, with kindly elders above her on the social ladder, low rank didn't matter. Misty saw how Pearl felt, and relaxed.

Nor was Sundog going to fall in love with Pearl, a possibility that seemed not to have occurred to anyone except perhaps the human beings. If he had posted a Keep Out sign for Pearl as he had done for Misty, Pearl didn't know it, because she had gone into the house without waiting to see what he'd do. A more stress-free introduction of a new dog would be hard to imagine, thanks entirely to the social sophistication of Pearl, which she had acquired in the Boulder household. Like a young, well-trained girl straight from finishing school, Pearl was graceful. Her confidence and good manners put other dogs at ease. She was, in short, socially secure.

Pearl must have been pleased at what she found in the kitchen. There were the cats, all of whom had come from our son's home in Boulder. Ordinarily the cats would have bolted when a new or strange dog entered the house, but this time they did not. Several people happened to watch this meeting, and expressed surprise, but few people appreciate the memory of animals—an ability which has been extensively studied and conclusively proved by scientists, incidentally—hence that the cats remembered Pearl was not unusual. Soon enough they greeted her, though not all at once, brushing her nose lightly with theirs and briefly arching their backs under her chin. Pearl received their greetings calmly, as if she and the cats had not been parted for any length of time, or as if she took their presence for granted.

Thus she fit our group as if she'd been born to it. And of course, in a sense, she had. My husband and I must not have seemed all that different from our son and his family, and anyway, she knew us from our visits, to say nothing of her early life among the cats. Meanwhile, she did not forget her natal group in Boulder, which was obviously very much in her heart and mind. We saw this the first night she spent at our house. In Boulder, the dogs had developed the habit—no one knew why—of eating lying down. Pearl had the habit also. In Boulder the dogs were fed at night, and when I fed our dogs that night, Pearl lay down to eat as was her custom. I found that charming, especially when, after a bite or two, she glanced quickly and questioningly at Sundog and Misty, who of course ate standing up. Pearl seemed to be waiting for them to lie down too. They didn't. Pearl glanced around at the rest of us, perhaps to see if we had noticed the strange behavior of Sundog and Misty. But evidently, we weren't reacting as expected. *Oh well,* Pearl seemed to say, and returned to her food.

Unlike the dogs in Boulder, who ate only in the evening, our dogs ate twice a day, so in the morning I again put three bowls of dog food on the floor. Pearl was always hungry, always ready to eat anything—she would snap up whatever dropped, including pills and bits of paper—but that morning she again held back from her food for a moment while she considered the other dogs, who had already begun to eat, and who of course were standing. *The elders show the way,* says the dog law, and Pearl obeyed. If Sundog and Misty ate breakfast on their feet, then so must she. She walked up to her bowl and ate from it standing.

Even so, she couldn't forget her elders in Boulder and all that she had learned from them, so when dinnertime came, she again lay down in front of her bowl. She kept this up for many months, standing up to eat breakfast, lying down to eat dinner, showing her fidelity to both sets of elders until the pull of her father and the other dogs in Boulder faded into the past.

10

ONE of the first qualities I noticed in Pearl was her excellence in the woods. Often, she and I went out together, and early on, I realized that she was very different from many other dogs, who normally course around, making their own investigations, associating with their people only now and then and only for the purpose of arriving at the same place later. Some dogs don't even seem to care that the woods are very different from human environments. To such dogs, an outing is an outing, nothing more. I've been in the woods with many a dog who wanted only to dash around in circles with a stick in his mouth.

Pearl, however, forgot about play when she was in the woods. She even forgot about other dogs. Rather, she stayed

right with me every moment, always alert for scats and footprints, carefully investigating every pile of deer droppings, every signpost marked by coyotes and foxes, always looking up into the trees and under the bushes, and always, when she noticed something, looking quickly up at me to learn if I had noticed too. Many things I might not have noticed, but thanks to her, I did. Why did she do this? I believe it was because she was acutely aware of other animals. Having been raised in my son's busy household, she had lived with many other animals—the dogs, the cats, the rabbit who hunted like a dog. Hence all animal activity, not just dog or human activity, seemed very important to her. It did to me too—hence we had much in common. I had felt very close to Pearl when she first chose me in Boulder. After a few trips through the woods with her, I felt I couldn't live without her. Thanks to her, I learned more in a week than I would have learned in a year on my own. She was my sister.

Pearl was a dog who made an art of barking. This was perhaps not surprising. Unlike wolves, who must communicate only with other wolves and therefore seldom need to bark, dogs must communicate with people, and we require cannonades of sound. We are dense where canids are sensitive, and dogs must exert themselves mightily to make their feelings known to us. Hence one of the most significant things a dog can do is bark, and Pearl knew it. I saw no point in trying to silence her, since she was so loud that she couldn't hear me anyway, and if I raised my voice, my shouting only made her bark the more. After all, acoustically speaking, a shout is like a bark, so shouts merely stimulate dogs. Stop! cries the person. *Raugh!* cries the dog. Stop! *Raugh!* Stop! *Raugh!* As the dog sees it, the person is joining in to repel the intruder.

Early in life Pearl had figured out that people respond well to this stimulus, so she would bark when she wanted food or water, when she wanted a treat, when she wanted to go in or out, when she wanted your attention. She would even bark while holding something in her mouth. In our area, the drivers of the delivery trucks often carry dog biscuits with them, and when the FedEx driver or the UPS driver would come to the house, he'd hand out biscuits. Pearl, of course, would be barking at him loudly, but she'd take the biscuit anyway and bark on with the biscuit clenched in her teeth. She barked at anything unusual, and at anything familiar. In short, she barked at everything, even at robins on the lawn. She'd look out the window and see them hunting worms. Our lawn! Swarming with intruders! She'd rush outside to chase them, barking. *Go, go, go, go!* her voice said. The robins would take to the air. She barked at all our friends, following them around the house, barking, and if they spent the night, she'd bark when they came downstairs to breakfast in the morning. One of our dearest friends and most frequent visitors was the Serbian writer Vladimir Pistalo, and he felt that Pearl was ethnically Serbian. She was, he said, a deeply suspicious dog.

Some people find barking obnoxious. Sometimes I did too, but thought I might as well try to learn from it, as I had no choice but to listen. In the end, I was glad I did, because I learned that Pearl barked in several different voices. She used a special bark when making requests—just one bark, not too loud. If the request wasn't honored, she'd wait briefly, then give a single bark again, politely refraining from barking more until she'd learned whether or not her first two barks had been effective. Thus a dog might bark to get the attention of another dog, especially if the other dog were at a distance.

However, when it came to barking at intruders, Pearl's voice was entirely different. What's more, she barked in different tones for different creatures. Amazingly, she used one tone for dogs, another, slightly different tone for large nondogs, and a third very different tone for small nondogs such as skunks, porcupines, and ravens. It came to me that because Pearl was a skilled observer in the woods, noticing every sign of other animals, it would be natural that she would make such distinctions. Through her different styles of barking we learned that to Pearl coyotes were dogs, and people were large nondogs, as were deer, cars, sheep, horses, herons, and the like. I was excited to learn that foxes were also nondogs. Just as I was thinking I'd never hear her bark in the presence of a fox, one came out of the woods, and she barked at it. A small nondog! Imagine that! Foxes belong to the same genus as dogs, but unlike coyotes are not in the same family. Pearl's views agreed with those of a zoologist.

Even so, she didn't welcome small nondogs, and when she found them in the field she let them know as much in an excited, high-pitched voice. *Don't come back here!* her bark said. Sometimes the wary little nondog would climb a tree, in which case Pearl would stand below it, barking upward. However, she had no wish to hunt, hurt, or kill little creatures—she never chased them in any meaningful way. She just wanted to alert them to the fact that they were trespassing. Or at least, that's what she did if the nondog was wild.

If, however, the nondog was a cat, she saw it as a threatening intruder just as she saw strange dogs and people. Many of the members of our group were cats, and evidently Pearl's proprietary sense included all group members, not simply the other dogs and the people. Once when a homeless cat came shyly across the field toward our house, perhaps looking for shelter, Pearl reacted instantly, mercilessly, rushing it, then chasing it at full speed, not for fun but to kill it, then diving at it, grabbing it, and tussling with it as it fought for its life. We managed to pull Pearl off and tried to catch the cat—we would always adopt a cat in need—but during the struggle, it escaped. We tried to find it, but understandably it had vanished in the woods. Fortunately, a few days later a cat matching its description showed up at a neighbor's, and was taken in.

As far as large nondogs were concerned, all posed a threat to our group-held territory. The very thought of deer seemed to bother Pearl, who often asked to go out after dark for no other reason than to look in the field in case the heinous animals were eating our grass. If they were, she would rush at them, her voice ringing. Some dogs run deer, a natural outcome of their wolfish hunting instincts, but Pearl simply wanted to chase them off the place. So that was all she ever did, rushing at them, barking, until they flung up their white tails and bounded off, relinquishing the field to Pearl. As soon as they vanished in the woods, she'd come back to the house feeling satisfied, her job done.

Barking is often significant, but not barking is significant too, or it is at times when barking might be appropriate. Sundog was very discretionary in his barking, for instance. Dogs such as he inspired the saying: "When the old dog barks, you'd better look out the window." And when we heard his voice, that was exactly what we did. But no one, not even the other dogs, would pay much attention to Pearl, or not when she was young. I once observed her barking at a housefly—if not with much energy, to be sure.

Yet even young Pearl sometimes found occasions for silence. Once a mountain lion came out of the woods. I happened to notice it through a window. "Oh my God!" I said aloud, and ran out for a better look. Pearl jumped up and ran out too. She had never seen a mountain lion before, and through my very excited mind flashed the notion that she might chase it. It was a strange cat and also a large nondog, either of which she would normally feel the need to threaten. Unlike the others, though, this one would kill her. Just as I reached out to grab her, she saw it. But instead of barking or trying to rush at it, she stopped dead in her tracks and stared transfixed, her eyes big, her ears forward, her body rigid and her hair bristling. Not a single yap escaped her. She too was saying, *Oh my God!*

On another occasion, one moonlit night, a visitor who was

spending the weekend suddenly appeared in a panic to tell me that a large man in dark clothes was standing near her bedroom window on the far side the house. This seemed improbable, but I couldn't dismiss the warning, so I ran out for a look.

Pearl wouldn't let me do this by myself, and ran out right beside me. Intelligently, knowing from the demeanor of the human beings that something serious was afoot, she didn't bark at all but waited to see what might develop. In silence we ran together around the house—she very tense, close to my side, and ready to fling herself at whatever it was—but we saw nothing. Nobody was there, and no car was on the road. Nevertheless, Pearl's nose went up, the hair on her back did too, and she suddenly stiffened as she had done for the mountain lion.

We were near the place where the large man had supposedly been standing. Nearby was a bush on which a grapevine grew. The vine was pulled loose, and one end was dangling. Clearly, the visitor had seen a bear, who must have been eating the grapes.

Pearl knew something of bears, although I'm not sure how. Bears would not have been around in Boulder, and in our travels through the woods, we had never met one. Yet that she knew about bears I had no doubt. One moonlit night in early spring, as often happened, I decided to take a walk on the road and started off down the driveway. Usually Pearl came too, with enthusiasm. But on this occasion we hadn't reached the road before I realized that Pearl was not beside me. I looked around and saw her sitting alone, partway down the driveway, watching me with anxiety. When I called her, she stood up but didn't come. Rather, she looked back at the house. When I started off again, she gave one bark to alert me. I turned again to look at her, and again she was sitting. She didn't want to go and she didn't want me to go. Reasoning that she knew something that I didn't, I deferred to her better judgment and turned back. We both went home, and the next day we found the tracks of a young bear in the snow. I would not have suspected a bear that early in the year, having thought they would be hibernating, but bears hibernate because they want to, not because they have to, and some do emerge early sometimes, as this one had done. In our area, no one has ever been harmed by a bear as far as I know, but dogs may have. Pearl was probably right to be careful.

On the lawn outside the guest room window, Pearl and I listened for something moving through the underbrush but we heard nothing—perhaps the bear was hiding nearby. Pearl's eyes scanned the dark bushes, but she made no sound. So it seemed that her worldview might include a fourth category—big, dangerous nondogs with teeth and claws that can rip a dog apart—and that her response to Category IV was dead silence. Alert but quiet, she followed me back inside, keeping between me and the bear, now and then looking back over her shoulder in case he should appear. Evidently he really was hiding nearby. In the morning, the entire vine was lying on the grass and all the grapes were gone.

Despite Pearl's many sensitivities, she was not sensitive about her own person. In this, she was less like a living, breathing organism, and more like a bowling ball. Indifferent to pain, and in fact to any hardship, she would charge around recklessly whenever she got excited, squeezing through spaces too small for her body, knocking things down without noticing. Once while on the porch, she evidently heard another dog barking and she suddenly wanted to get outside fast. However, rather than go back through the house and out the dog-door, she tried to get out through a cat flap in the screen. Her head went through, or part of it did, but the opening was much too small for the rest of her. But that wasn't important—she went through anyway, oblivious to the ripping screen and the gaping hole, like the exit wound of an artillery shell, that her determination caused.

Once when we were walking in the woods Pearl found a small burrow in the earth, sniffed at it with interest, and began to dig. Out came a swarm of white-faced hornets, who stung her fiercely. Shaking her head to get rid of them, she ran straight to me, sat down squarely on top of my feet, and began biting the hornets, who, interestingly, concentrated on her and left me alone. Hornets don't lose their stings as bees do—each hornet she bit would sting again inside her mouth. She'd spit it out and bite another, dealing with them one by one as dozens stung her. I lost courage and ran, calling desperately to Pearl to follow, but she was too busy to obey. At last she realized that the hornets were stinging her much faster than she could bite them, so she stood up and shook herself furiously. Even that was useless. Head down, tail drooping, she tried to walk away.

But the cloud of hornets followed. Pearl seemed to ignore them. Her skin twitched with each new sting, but from her deliberate manner no one would know that she was suffering. Since she was forcing herself to walk rather than breaking into a run, as she must have been longing to do, she seemed to be staging this falsehood for the benefit of the hornets. An age-old rule of the animal kingdom says not to run if your adversary is within striking distance. Animals obey this rule, and so do people if they're smart. Stay calm. Be cool. Pretend you're not worried. The last thing you want is for the enemy to chase you. When Pearl was a certain distance from the hornets' burrow, they returned to it, leaving Pearl covered with lumps and moving stiffly. But as we hurried home for baking soda, she never cried or complained.

One summer night I heard her barking in a voice that said she'd found a small nondog. That and the fact that she stayed in one place meant that she'd almost surely found a skunk. (Skunks are marvelously unafraid, and stand their ground no matter what.) Hoping to save the situation, I went out with a flashlight. The barking stopped. In the dark field I saw two pairs of eyes, Pearl's large and green and high off the ground, the skunk's right beside hers, small and red, down in the grass. Both Pearl and the skunk had turned to look at me. I called to Pearl.

The eyes vanished as she and the skunk faced off again. Pearl's voice rose in speed and pitch, then very suddenly stopped in mid-bark. I heard a cry, then dead silence, then a cough. The beam of the flashlight showed Pearl staggering backward. She'd been sprayed full blast in the face.

Tear gas is better than the burning, choking spray of a skunk at very close range. Pearl didn't seem to be breathing. I ran toward her. Gasping for air, she slowly turned around and walked away, head down, tail drooping, eyes squeezed shut, not toward me but toward the house, stopping often to bend her head and rub her eyes against her front legs.

At home, I had nothing to remove the devastating odor, so, having heard that a certain brand of douche takes skunk off a dog, I had no choice but to rush to the local pharmacy and request twelve bottles of the product. Twelve bottles of douche? Why would anyone need so much? All the other shoppers turned to look. But if Pearl was tough, then I could be too. Bearing the enormous load of douche I left the store with as much dignity as I could muster. And yes, the douche worked. Pearl soon smelled better.

Naturally, Pearl assumed that others were as tough as she, so she was always injuring someone. Interestingly, however, she confined her roughness to people. With the cats of our household, and also with the other dogs, she was unfailingly gentle. I would have preferred that she be gentle with people too, but there wasn't much I could do about it. No matter how much I yelled at her to stop, when she jumped up on me in greeting, as she would after even the shortest absence, her claws dug bleeding furrows down my skin. Nothing deterred her, not even the shock-collar which she, like Misty, briefly wore to keep her off the road. One day I wanted to walk to a neighbor's house. Pearl wasn't invited, so I put her collar on to stop her from following me. She knew about the shock, and knew she was wearing her collar, but she followed me anyway until she heard the warning buzz. It had stopped her before, but not this time—she stood still for a moment to gather her courage, then, ears low and eyes squinting in anticipation of the shock, she suddenly rushed at me, getting a dreadful jolt that forced a little cry from her throat, but not stopping until she was on the road, greeting me as if we'd been apart for years.

Not many of us will brave an electric shock, not for any reason. Pearl was different. Interestingly, she did the same thing again on another occasion, so there was no mistaking her courage. Since the collar didn't work for Pearl, we took it off, and didn't have further chances to observe her brave behavior, nor did we want to see it, but even so, I couldn't help but notice that she'd dealt with the problem just as a person would have done. After all, a shock is a shock, and feels the same whether you are walking or running when you get it. Still, for psychological reasons, a person would also stand still for a moment to gather courage to take the shock, and then run fast to get through it.

Pearl's toughness could be daunting. One day she was following me to the car, fully aware from the clothes I was wearing and the suitcase I carried that I was about to leave home but would not be taking her with me. This saddened her, so as we both walked slowly up the driveway together, she pressed her nose against the back of my leg. Suddenly, ahead of us, a squirrel dashed across the driveway. Pearl hurled herself at him, bowling right under me and knocking me into the air like a ninepin. Back to earth I fell and broke my ankle.

On another occasion she was closely following me down a flight of narrow stairs when she heard a man's voice at the kitchen door. Perceiving a threat to the household, she yelled and charged him, slamming into the back of my legs and knocking me forward. Straight to the lower floor I plunged, and almost broke my neck.

Still, I continued to adore her. I liked her strength and her self-reliance. And I liked her good intentions. Whatever she did, however roughly, she did with others in mind. For the good of her group, she challenged strangers. For the sake of joining me, she braved the electric collar. If she raked furrows in my skin from jumping on me, she meant only to greet me. If she knocked me down once or twice, it was only because she was trying to be near me when, by chance, an urgent matter called her. When she later found me on the floor, she grew wild with concern and climbed all over me, kissing me profusely and digging at me with her powerful claws. Obviously something terrible had happened to me. Who knew what or why? She didn't want me to be hurt. She didn't want me to cry. She urgently did what she possibly could to help me feel better immediately. Who could resist such a dog?

Early in our relationship, I gave Pearl the towel-biscuit test, assuming she'd pass easily. I put the biscuit on the floor and covered it with the towel, but to my surprise instead of going after it she stood looking up at me. "Go ahead," I encouraged her. "Take it." Still she looked up doubtfully.

I waited a while, got bored when she did nothing, remembered something in another room and left Pearl alone briefly. I came back almost immediately, but found the towel crumpled and the biscuit gone. In my absence, Pearl had snatched it and bolted it. She seemed startled to see me, and perhaps a little guilty. Had she passed the test or not?

These tests don't account for the fact that dogs sometimes see things differently than we do. Perhaps Pearl thought that if I wanted her to have the biscuit, I would have handed it to her. So perhaps I wanted it for myself. Dogs hide food by burying it. Hadn't I more or less buried the biscuit by covering it with the towel? To her, I suspect, my actions seemed ambiguous, at least ambiguous enough to deter her from taking the biscuit while I was in the room. But the moment I left, the situation clarified. If a dog finds food abandoned by another dog, she eats it—no ambiguity there.

When Sundog took the test, I was standing beside him, smiling and expectant. He knew all about training, all about people waiting for him to do the right thing, and all about rewards. Clearly I meant for him to take the biscuit. If he had seen the problem as Pearl saw it, he wouldn't have touched the biscuit, as he proved to us time and again when left alone with human food. But then, human beings had taught Sundog how to conduct himself. Pearl was different. Pearl had learned her ways from other dogs.

Hence, only someone who would confuse obedience with intelligence would feel that Pearl had failed the test. In fact, she was very intelligent, a quality she displayed time and again, never more clearly than she did when away from home, in circumstances that were to her unfamiliar. I had been invited to a bookstore in Providence to sign copies of a book I'd written, and, since the book was about dogs, I was asked to bring a dog with me. Which dog to bring? Who better than Pearl! She always wanted to go where I went, and when she saw she was included, she was happy. At the car I opened the rear door for her as I usually do, and she jumped in, but while I was getting in on the driver's side she jumped over the seat so she could sit beside me. There she could be close while keeping a lookout for strange, oncoming cars and for strange pedestrians at the roadside, and warning them away by barking.

The nearer we got to the city, the more cars and people she saw. Eventually she was barking all the time, first at one stranger, then at another, until she felt she couldn't cope with the horde of interlopers and got down on the floor. She knew they were there, though, and soon she was up on the seat again, pressing close against me and watching the passing cars suspiciously through the window.

Trying to keep my wits about me had been difficult with all the barking, and unfortunately I got lost as soon as I left the thruway. I then became disoriented in the maze of one-way streets. The time to start the book signing came and went while I was trying to find the address, then more time passed while I looked for a place to park. But there are no empty parking spaces in Providence. Fifteen minutes passed, then half an hour while I drove with increasing panic up and down the parked-up streets. At last I found a space into which I managed to squeeze the car, and with Pearl hot on my heels I dashed for the distant bookstore, asking directions on the fly.

In the store I signed books and talked with the customers while Pearl crouched under the counter in an effort to escape the crowds of book-buying dog lovers whom she didn't even know yet who kept patting her anyway. If the overpopulated highway had been bad, the reaching hands of a whole new group of strangers was much worse. She hated everything about the signing and kept looking at me to learn if we could go. When finally the time came, she was ready. She shot out the door ahead of me, happy at last.

But where was the car? I had never before visited that section of Providence, and had no idea where I had parked. Nor, in my panic, had I noticed the name of the street I had parked on, or the route I had taken to reach the store.

Now what? If I'd lost the car in a parking lot, I would have resigned myself to waiting for night to come and the lot to clear of all cars but mine. But this was the city. Only the threat of nuclear attack would clear the cars out of the city. Failing that, I had no idea how I was going to get home. I couldn't expect the police to find my car for me because it was of a very common kind and color, and I didn't remember the number of the license plate. Deeply concerned, I looked down at Pearl. I might be able to find a hotel room for myself, but what would I do with her if I couldn't find the car?

Meanwhile, Pearl was looking at me, as if to learn why I was stalling. Could I possibly want to go back inside the dreadful store? *Enough of this,* she seemed to think. *We're getting nowhere. I'll take charge.*

And she did. Lowering her head, she gave an overpowering tug at the leash that yanked me into the street. She didn't look back to see if I approved of this—she just pulled me forward. Nor did she check with me after we crossed the street. Instead, she tugged me onto the sidewalk and, puffing with determination, pulled me rapidly down the block, then down another block, then another and another, turning left, turning right, until I had no idea where we were. Still, she seemed so very sure of herself that I ran along anyway, led by her upturned tail and her solid hindquarters.

She kept to the sidewalks, narrow though they were, taking to the streets only to cross them. People passed us but she didn't bark; dog scats and urine stains were everywhere, but she ignored them. A tremendous purpose radiated from her hurrying body. Onward she plowed.

Every street was parked solid on both sides. The bumpers of many cars were touching. Cars were parked across driveways and next to fire hydrants. Trying to scan them all in case one should be mine but seeing none that looked familiar, I realized we hadn't been on those particular streets before. We might be following the cold trail of a traveling squirrel, or (and I was chilled by the thought, as I had heard of dogs homing) we might be heading on foot for New Hampshire.

No more of this, I decided. Better to go back to the bookstore and start over. But I couldn't remember its address. Worse yet, I had forgotten its name. Would I recognize it if I saw it in the Yellow Pages? I looked in vain for a phone booth and was just about to jerk Pearl to a halt so I could go into a shop and ask the shopkeeper, when abruptly she stepped off the curb. I thought we were going to cross, but instead, head down, she purposefully turned the corner and churned along the middle of the street. An oncoming car had to pull over to make way for us. Pearl ignored it. Just as I was wondering why we were not on the sidewalk, she stopped so quickly that I almost fell over her. Briefly, she glanced up. Our eyes met. Her manner said, *Okay, that's it.* And so it was. Beside us was my car. And not just that—we were by the driver's door.

We could go home. We wouldn't have to spend the night behind an ash can. Half smothered by my almost tearful praise, Pearl remained unusually calm, not responding with jumps and little yips of pleasure, as she ordinarily would do. *Yes, here's the car,* she seemed to say, *but it's up to you to drive it. Why are we standing in the street?*

We got in, she first. When the car started, she happily rolled out her tongue. When I finally was able to stop telling her how great she was—an outpouring which she again took calmly—I began to wonder how she had achieved this miracle.

When animals solve problems, we human beings, self-important creatures that we are, seldom recognize what they're up to, let alone see how they do it. Our powers of observation don't approach theirs, and furthermore, the problems they solve, however difficult, often seem trivial and thus pass beneath our notice.

My own priorities were a perfect example—because I'd been lost, I'd been overwhelmed by the fact that Pearl had known all along where the car was, so I almost overlooked the rest of what she'd managed. Hence I spent the first part of the drive marveling at her sense of direction. I felt sure she hadn't retraced our steps to find the car. Thus she hadn't followed our scent trail, if, indeed, it hadn't already been obliterated. Rather, as if her own personal satellite hung above her in the sky, beaming down the appropriate coordinates, she'd taken a direct route.

But many animals are excellent at homing, as I already knew. Her innate skill should not have surprised me. I should have taken note of the many other things she'd done that were not innate and therefore more remarkable.

For instance, how did she know to take charge? Had she understood my helplessness and confusion? Or had she simply felt that I was being slow, so that her impatience impelled her?

And how did she know to walk on the sidewalk? At home, there were no sidewalks, so we always walked on roads. Still, on our way to the book signing, we had walked on sidewalks. Was she copying my method? Or did she realize that city streets were dangerous? And if so, wasn't that quite a mental leap for an inexperienced country dog?

But if she realized the danger, why then, when near our car, did she deliberately step into the street in front of an oncoming car? Did she think I wouldn't be able to squeeze between the tightly parked cars? She would have gone through easily enough, but I would have had to climb over the bumpers. Or did she want to approach the car on the driver's side, by the door I always used?

And why did she take such pains to approach that particular door and no other? At home, we always opened the rear doors for the dogs. They virtually never used the driver's door. They knew this, and would wait by a rear door when they expected to go for a ride. What was different about this occasion? We had both gotten out the driver's door. Did Pearl see our journey as a process which she wanted to reverse? Or did she see the driver's seat as the command position, and, being at the time in complete command, decide to enter the car accordingly? Or was she thinking of the drive itself, and therefore the driver, and therefore the driver's seat and the driver's door? Or did she simply hate Providence so much that seconds counted? Did she feel that if we both used the same door, we'd get home faster?

Whatever the answers, whatever the reasons, there was no denying that Pearl had done a remarkable thing. I kept glancing at her with admiration. She caught my eye, and, turning her ears forward, she shot me a sharp and meaningful look. I felt as though I'd been reprimanded, told to focus, to watch the road, to pay attention, to get us home. So, obediently, I looked at the road. Humbled by this dog's ability, chastened by my own ineptitude, I didn't feel at all sure that I was the dominant creature. Rather, I felt dependent, just as a dog might feel when its master is in charge.

Even so, I couldn't stop thinking, and the more I thought, the more sure I felt that Pearl had perceived my bewilderment, and understood that if she didn't force the issue and drag me to the car, she'd never get home. I also felt reasonably sure that she brought me to the driver's door because it was the door I always used. Maybe she thought I wouldn't be able to find it if we approached the car from the sidewalk.

11

A FEW months after the experience in Providence, I was again invited to speak on the subject of animals, this time by a cable television station in southern Vermont. The program was produced in cooperation with a local animal shelter, which used its airtime to display some of the dogs and cats who were available for adoption. My task was to discuss these prospective pets, pointing out their good qualities in hopes that people in the audience would give them homes.

The Green Room was the cattery, where I became fascinated by a certain white cat. An intact male four or five months old with a few black spots and a long black tail, he had more poise than virtually any other animal I'd ever seen. Serene although trapped

in a cage, confident although imprisoned in a room with ten or twelve other cats who were by no means serene or confident, he seemed unfazed by his situation. He saw me watching him, but didn't seem bothered. Rather, he met my gaze for a moment, then turned his eyes toward something else, as if to him all things were equal. Many a cat would crouch uncertainly to watch the watcher. Not he. Obviously, whatever fate might have in store for him he thought he'd be able to handle by himself. Meanwhile, he had used his time in the cage to put his fur in order. Perhaps he was homeless and imprisoned, but he'd kept himself beautifully clean.

The woman who ran the shelter came in, and I asked her about this white cat. A stray, he'd been picked up a few weeks earlier, she told me. No one had looked for him—evidently he'd been abandoned. This was in October, after the summer people had gone home. In Vermont and New Hampshire, many homeless cats appear on the streets and roadsides at this season. The woman's guess, and mine too, was that summer people had taken the white cat as a kitten so that they could have a pet during their vacation, and then had tossed him out when their vacation ended and they went home to their pet-free apartment building.

I asked to see the cat. The woman opened his cage. I lifted him out. He considered me briefly, then abruptly rubbed his chin on mine, a quick rub to the right, then a quick rub to the left, although I was a perfect stranger. *Your kind is all the same to me,* his manner said, *but I rather like people.* It was as if he'd known me all his life.

During the program, I held up the white cat and said very good things about him. I also did the same for the other cats—all but two fragile, frightened kittens who crouched together in the back of their cage. As distressed and confused as the white cat was confident, these kittens had been found in the woods in a hollow log. No use holding them up in front of the camera—they would have struggled and scratched and made a bad impression.

After we had displayed the other cats to the viewing audience, someone brought in a dog who had been housed in another building, the only dog on the premises at the time. Weighing about forty pounds, she looked something like a German shepherd, but much smaller. I wondered if she too had been a stray, but no, she'd been given up because of behavioral problems. She wouldn't come when called, she leaked urine so couldn't be completely house-trained, and she made a terrible sport of killing cats and chickens.

Perhaps she was a bad dog, and perhaps she leaked urine, but she was overjoyed to be out of her cage and in the room with us. Smiling, she immediately sat down and opened her hind legs to show us her belly. I thought her posture a bit unusual—most dogs lie down for this display. Nevertheless, anyone could see that she was a polite dog who wanted us to like her. Obviously, she was trying her best to charm us with her pleasant ways as quickly as she could, as if she realized that she wouldn't have much time to do it.

During the program I said many good things about her too, looking down at her and pretending to talk to her to make her smile. She did—she was very responsive—and when I leaned down to pat her, she curled herself around my hand and wagged her whole body. When the program was over and one of the producers took her leash to lead her from the room, she gave herself to him in ecstasy, rejoicing with graceful, dancing leaps and little yells of pleasure. Clearly she thought he wanted her and would be her new owner. Since he was actually taking her back to her cage, everyone present felt a bit sorry for her. She may have noticed our expressions, but she didn't know what they meant until she saw for herself where she was going. And then, as if a light went out, her joy vanished, and she began to cry.

That did it. I took her. I took the white cat too. And I also took the two feral kittens. Thus we increased by four the population of our mixed-species group.

My husband later named the white cat Pae—pronounced Pie—as in *paesano* (actually, *compaesano*), because of his confidence and his bold but friendly manner. My husband also named the feral kittens Wicca and Phyllis Sylvestris, Wicca because she was black and could have been a witch's cat, and Phyllis because she looked exactly like her tabby relatives, the wildcats, *Felis sylvestris*. The dog already had a name—Ruby.

All went well on the ride home—Ruby curled happily against me—but her first act upon leaving the car was to streak off after Rajah, Male Cat One, who, tail high, had strolled out to greet me. Rajah was horrified. Completely unaccustomed to dogs who would chase him, he bolted up a tree and spat at Ruby. She didn't care—she'd already noticed one of the other cats and shot off after her. That cat made it through the dog-door just one leap ahead of Ruby, who then noticed a third cat and sped away in pursuit.

Sometimes when a dog chases a cat, the dog doesn't run her fastest even though she seems to. Instead, she keeps a short distance behind, slowing down slightly if she's gaining too fast. This is because she doesn't mean to kill the cat. Many animals learn chasing and killing as two separate, almost unrelated skills, and therefore, in some chases, actual capture is not the object. The longer the dog can keep the cat running the more fun the dog has.

But Ruby gained alarmingly on our cats, showing that she understood the entire process, that both the chase and the kill were amusing. No use screaming at her to leave the cats alone—this was the behavior that put her in the shelter in the first place. Within moments, she had treed three cats and driven off two others, and was looking around for more.

In our household at the time, we not only had cats but also parrots. I was beginning to question the wisdom of bringing an uncontrollable killer of cats and chickens into this particular ménage, wondering why I hadn't foreseen some of the problems earlier, when Sundog came out the dog-door, followed by Misty and Pearl. Realizing that a strange dog was present, Sundog bristled and lowered his head. So did Misty. Here was an interloper who might very well want Place Two.

Pearl simply looked on, interested but unworried, but since she was Dog Three, nobody paid much attention to her.

Ruby hadn't bargained for Sundog and Misty. As the two of them advanced on her, she threw herself flat on the ground, lay rigid, and lifted a hind leg very slightly. Not for them the charming, upright display she had shown to the people during the TV program. These were dogs, and they meant business. Better to display to them in the most doggish manner possible. Moving slowly, the two dogs grimly sniffed her all over, pointed their noses down at her to make sure she didn't think she outranked them, and marked here and there to display their status while Ruby cringed. Even her gaze was rigid. She made sure to look past them, not at them.

Meanwhile in the tree, Rajah had been watching. Noting that Ruby seemed under control, he slid backward down the trunk, turned around to jump the last few feet, and continued on his way. By then, the other dogs were done with their investigation. Sundog left, Pearl watched, and Misty walked around, stiff-legged, her eyes on Ruby's cringing form. Given the uncertainties of the moment, I thought a fight could start, so I called the three dogs indoors again, whereupon Ruby leaped to her feet and ran after Rajah.

But Rajah had had his fill. He turned to face her. She rushed up to him, blind to the fact that his back was arched, his eyes were blazing, and his fur was on end. Before she knew it, he was swarming all over her, dealing her lightning blows. Ruby screamed the piercing *Yi yi yi yi* of a dog hit by a car and tried to get away from Rajah. But Rajah wasn't finished. His rage had come, and he couldn't get enough of clawing Ruby. In her struggle to escape she knocked him aside, but he still wasn't finished and he chased her. She headed for the car but the doors were shut, so she swerved and sped across the field.

Cats don't have the stamina of dogs when running, and Rajah knew it. He slowed down and glared at her retreating figure. Then he boldly turned his back on her, and, ears low, fur still bristling, he went swaggering toward the house, for all the world like an ex-marine who has just won a fight in a barroom. As he walked he growled and spat, growled and spat, as if he were swearing under his breath. He was reliving the experience.

It would be hard to imagine a worse introduction of a new dog to a household. We had to chase after Ruby, who no longer wanted anything to do with us, and of course wouldn't come when we called her. We finally caught her and put her in a room alone with food and a drink of water, a room where she could calm herself and have something to eat while we gathered our wits and tried to think what could be done about her. The other dogs sniffed at the crack under the door, but Ruby's display must have mollified them—they seemed curious, to be sure, but they, at least, didn't seem to view her as a threat.

Meanwhile, all the cats had vanished. I thought we'd never see them again. But during the night they stole back to the house and hid in the basement and the closets and under the beds. In the morning, while we were feeding Ruby her breakfast, one of the cats happened to enter the kitchen, froze when she saw Ruby, then apparently decided that Ruby seemed harmless, and jumped up on the counter to eat her own breakfast. What had changed? We looked at the cat, and then at Ruby.

Ruby hadn't moved. She may have stolen a glance at the cat, but she kept her face averted. She didn't chase that cat, or any of the others who came into the kitchen later, or any of the cats she saw the next day. And as for Rajah, who obviously was still very much in charge, we began to notice that he was paying almost no attention to Ruby, and that Ruby was humility itself in his presence. If he jumped up on a chair where she was resting, she'd get off and give him her place. If both of them approached a doorway together, she'd hold back so he could go through first. I'm sure he could have taken her food if he'd wanted it. We'd been aghast at Ruby's behavioral problems with cats, never dreaming that Rajah would solve them for us. But thanks to him, she never chased a cat again.

As Rajah taught Ruby not to chase cats, so our blue and gold macaw, Rima, taught her not to harass birds. I had acquired Rima when she was young—a fledgling, really—and had named her for the ethereal bird-girl of the Amazon rain forest, the character played by Audrey Hepburn in the film *Green Mansions*. But Rima hadn't grown into her name. On the contrary, she became a ferocious, brawny creature who bit like a bear trap and talked like a sailor. If the Audrey Hepburn character were anything like my Rima she would never have perished helplessly at the hands of poachers—she would have grabbed an AK and smoked the fuckers (the macaw's vocabulary, not mine).

When Rima first saw Ruby, the pupils of her eyes shrank to pinpoints. This dog was trouble, Rima knew it, and raising her feathers to make her big self bigger still, she mantled her wings, opened her enormous beak, and yelled like a fire alarm.

Ruby tucked her tail and ran. Obviously, the bird was dangerous. After that, Rima or any other parrot could fly around the room without Ruby taking any notice, and could even walk across the floor in front of the former chicken-killer while the little dog, with lowered eyes and folded ears, pretended that no one was there. That Ruby recognized the parrots as potential group members, or at least that she no longer thought of them as potential prey, became clear later, when she gained enough courage to try, very politely, to sniff under the tail of the African grey parrot, Pilgrim, as he walked across the floor in front of her, just as she might sniff the anus of another dog. Birds don't use this method of fact-finding, and Pilgrim seemed affronted. His pupils tiny, he spun around and threatened Ruby with raised wings and open beak. *Cut it out,* his manner said. The threat was mild—Pilgrim was afraid of nothing and didn't hesitate to bite when he was displeased—but Ruby quickly withdrew her nose. Mollified, Pilgrim strolled on with folded wings, like a man with his hands in his pockets.

Thus the problems of Ruby's sport-killing were solved. This left the problems of her incontinence and disobedience. She knew about relieving herself outdoors—she'd ask to go out just like any other dog—and she seemed embarrassed when she leaked, which never happened when she was up and about, but only when she was relaxed or asleep. When an accident occurred, she would move away from the place, or even leave the room, head and tail low, ears down. She even seemed to try to prevent the leaks, and would often find a place to sleep where she could lie on her back with her hindquarters up on something, higher than the rest of her, as if she saw her pelvis as a bowl from which urine could spill. We took her to the vet. There we learned that when she had been spayed, the surgery had damaged her urinary tract. The problem was helped by medication, and also by waterproofing her bed with disposable absorbent pads such as those used in nursing homes. I put these between her mattress and its cover, and then, if she leaked, I could wash the cover and change the pads.

So she didn't leak urine to speak of, and what she did we could deal with, and she didn't chase cats or harm birds, but she still stole food from the table, raided the trash, ran away regularly, and wouldn't come when called. One day when I went shopping, she wanted to come with me in the car. I welcomed her, but when we reached the parking lot of the supermarket, she jumped out the open window, danced off into the sunlight, and wouldn't come back or let me catch up with her. Fortunately, she ran to greet a passerby, who then caught her for me, otherwise she might be roaming still. The passerby was astonished by her disobedience. He'd never seen a dog so willful, he said.

Neither had I. She knew about coming, but she just wouldn't do it. Even the cats were more cooperative than Ruby. Most of them would come when called and all of them knew to come in at night as we had taught them to do because of danger from coyotes. Touchingly, some of them even came during our useless bouts of calling Ruby. I'd be standing outside in inclement weather having yelled until I was hoarse, when suddenly a cat would present herself. What had brought the cat to my call? Probably the stress in my voice. I could picture the cat crouched under a bush somewhere, listening. *That's Liz,* the cat would think. *She's calling.* The cat would listen a bit longer. *She isn't calling me,* the cat would decide. But at last the cat would begin to wonder. *Maybe she* is *calling me.* More time would pass, and more yelling, with the cat considering it all. *It could be me. I'd better go,* she'd tell herself at last, and she'd come—the angel. But not Ruby, who would know perfectly well whom I was calling all along.

This failure of Ruby's was particularly distressing at night, when we would let the dogs out for the last time before going to bed. Knowing the drill, the other dogs would relieve themselves quickly and come back to the house, but Ruby would often vanish in the dark. No use to call her—even if she could hear us, our wishes meant little to her, and she wouldn't return. The only remedy was to walk her on a leash, but even this didn't always work, as she would anticipate bedtime and depart in advance. We tried leaving her out all night to teach her a lesson, but she began to bark at 2:00 A.M. and woke everybody up. So on the nights when she deliberately stayed away from the house, one of us had to stay awake until she got cold or tired and decided to come home.

It was all very embarrassing. If we had somehow managed to train a large number of cats, why couldn't we train this one little dog? We were losing sleep from waiting for her, and couldn't leave the dog-door open for her. The cats came in at night but they didn't know why, and if the dog-door was open they'd go out again. Nor did scolding improve Ruby's behavior. In fact, I assumed that her former owners had called her to them in order to punish her, making her less likely to obey, and also immune to disapproval. As we waited for Ruby in the evenings, my husband and I would look at each other, not at all sure what best to do.

12

TRAINING an animal is reasonably easy, or it is once the trainer understands how the training sessions seem to the animal. Animal trainers know this, of course, and have developed a method of training one another so that a neophyte can experience how the animal feels—a process to be recommended highly. In such a training session, one person takes the role of the trainer and the other the role of the animal. No words are used in the process, and no punishment either, only rewards, which in a session involving a human subject can be blasts of a whistle. (A real animal might require something more—perhaps a few edible treats given at random to make a training session fun and rewarding in itself, although many dogs are satisfied with praise.) A task is determined by the

trainer which the "animal" knows nothing about—the task might be that the "animal" puts her left foot up on a certain chair, say—the important thing being that task is meaningless since most tasks devised by trainers will at first seem meaningless to real animals.

When the session begins, the trainer should be waiting in a designated room, ready to blow the whistle the moment the "animal" happens to do something that leads toward the completed task. The "animal" wanders vaguely into the room, with no clue as to what the trainer wants. When she strays near the chair, she hears the whistle. When she moves her left foot, she again hears the whistle. She begins to catch on—he wants me over here, he wants me to move my left foot—ah! to *raise* my left foot—until gradually the behavior is shaped.

I had been shown the technique by a friend who trained animals for biological research, but I failed to recognize it later when I myself was trained by Pearl. Pearl wanted me to get up early in the morning, not at six-thirty or seven, as I would have preferred, but at half past four. And several months later, that's just what I was doing. It didn't come to me how this had happened until I began to review the events of the past few months, and then I realized that Pearl had subtly but efficiently trained me.

The process was unnoticeable. Long before dawn, Pearl would begin to think about her breakfast. At first, she'd wait until I got up, and then she'd rejoice. But in time, she began to push me with her nose very gently. This would wake me. She seemed to realize this, and then, if I didn't get up, she'd gently push again. If I still didn't get up, she'd push a third time and might also bark—just one woof, but full of meaning. I didn't want her to be barking because she'd wake the household, so at that point I'd tell myself that it was almost morning anyway, and I'd get up and get dressed.

Presently I was getting up at six. I didn't often look at the clock—I just got up and got going—but one morning as I put

Pearl's bowl of breakfast down in front of her I happened to glancc at my watch and noticed that it was already six and I had been up for quite a while. Somehow, I was waking up earlier. Never really knowing why, I soon found myself waking at five-thirty, and then at five-fifteen. Stumbling around in the dark doesn't start a day well, so when I began waking at five, I tried to regain control of my schedule. "Please don't, Pearl," I'd say when she pushed me. She'd wait a few moments, then gently push again. What could I do? Wondering if she were going to bark, I'd wait, eyes shut. She'd bark. Too wide awake to go back to sleep, I'd get up and, yes, I'd give her breakfast.

When I found myself waking at quarter to five I shut Pearl in the kitchen. She seemed to take this in good part, and that night all went well until some predawn hour when she began to bark as if for an intruder. Hearing her, the dogs in our bedroom barked too. When they stopped, Pearl barked anew and stirred them up again. No one in the house could sleep with that racket, so I got up. My tasks of the morning are to make the coffee, empty the dishwasher, start the laundry, write out the shopping list, prepare the birds' breakfast, and feed the dogs and cats. As I groggily put Pearl's breakfast on the floor, I saw that it was almost five o'clock. I was still too sleepy to realize that I'd been up for twenty minutes.

At this point, however, our mornings stabilized. Four-thirty was when Pearl wanted to get up. She did so from then on, and so did I. Dogs and also cats have a frighteningly good sense of time as we know it—not only by the sun and the seasons but also by the calendar and the clock—and Pearl wanted to get up when the big hand was on six and the little hand was halfway to five, every day, winter and summer, no matter how near or far the sunrise. So from then on, the animals and I all got up at that hour while the rest of the household slept. My behavior had been shaped, my training was complete, and my habits were established. From then on, Pearl ate breakfast at the time she felt was the right time. Four-thirty had been her goal. And actually, I also benefited. We don't always need as much sleep as we think we do, and early morning is an excellent time to work. Three cups of coffee and five hours of sleep are almost as good as eight hours of sleep, or they are for a while, anyway.

I began to apply to Ruby the training methods Pearl had used on me. I would call Ruby, and then praise her lavishly if she as much as looked in my direction. She didn't expect praise—far otherwise—and she would seem surprised and faintly pleased to hear it. Occasionally she would even start to move hesitantly in my direction with a polite expression on her face. I thought we were getting somewhere. However, to train an animal is one thing, but to retrain an animal who associates obedience with misery and punishment is quite another thing. The process promised to be lengthy. That in itself was not discouraging—after all, it had taken Pearl at least four months to shape my behavior to her liking—but I knew that Ruby would be in danger as we walked on the roads if she wouldn't obey when we called her.

One day, Sundog happened to pass by just as I was calling Ruby. She didn't come but he did, even though he knew it wasn't he whom I was calling. Still, I think it bothered him to hear me calling in vain, and perhaps he wanted to show me that my wishes meant something to him if not to Ruby. Anyway, he slowly approached me, waving his tail politely and holding his head low. What a dog! Because he and I understood each other, I thanked him quietly, not in the exaggerated terms I had been using for Ruby. Ruby had been moving off, as if she hadn't been thinking of coming to my call, but she stopped when she saw Sundog and turned as if to come to me after all. She didn't—she changed her mind—but Sundog seemed to have made an impression on her. The next time I called her she hesitated for a while before moving off, and the next time I called Sundog and he came, she came too. Although dogs do indeed learn certain things by following the example of other dogs, obeying human commands isn't one of them, so when Ruby seemed to copy Sundog, she probably hadn't said to herself, *Oh, so that's what's wanted of me.* Rather, she was probably remembering the important dog law that says *The elders show the way.* Dog One cooperated with human beings. That being true, then perhaps so should she.

For all the difficulty, however, I never doubted that Ruby was trainable, bad as she seemed. All dogs are trainable, and sooner or later most low-ranking dogs will learn at least some skills from high-ranking dogs whether the owner helps in the process or not. All that is required of the owner for further training is that he or she not botch things up by making the process disagreeable, something that is easily achieved if only the owner can convey to the dog that the new skill is something that any dog can acquire, or else is something that the dog and the owner will do together, and that ultimately everyone will be happy and pleased. If, on the other hand, the owner makes the dog feel disapproved of, insecure, and always in the wrong, the dog will lose heart and get the impression that the owner can never be pleased. Hence I started on an optimistic note with Ruby, the only question being how long the process would take. After all, Ruby had been raised by people who probably had as little understanding of dogs as they had of training methods—people who evidently had assumed that punishment, probably harsh punishment, would eventually pay off. When it didn't, they got rid of the dog.

But even an emotionally damaged dog can learn if conditions are favorable. In time, Ruby began to come when called just as our other dogs would, unless some compelling reason prevented her. Perhaps a professional dog trainer wouldn't have seen sufficient improvement, but I did. Even Sundog didn't come when called if something more important made him want not to, a behavior it had taken months, even years, of relaxation to instill in him—a behavior which made him even more interesting than he was at first, when he was sensitive to our every whim, as his former owners must have taught him to be. So the difference between Ruby and Sundog in this question was merely that Ruby was less attuned to my wishes than Sundog was. Even so, she was emulating Sundog, a sign that she felt she belonged.

Still, her responses remained interpretive. At night, for instance, the dogs liked to sleep in the room with me and my husband. So at bedtime we would call them from whatever they happened to be doing so that they, accompanied by many of the cats, could follow us upstairs. And every night, Sundog, Misty, and Pearl would do just that. Not Ruby. Knowing that we were about to retire, she would have gone quietly upstairs ahead of time, often creeping up the back stairs in order to be inconspicuous. In the bedroom, she would have taken the best and most popular of the dog-beds.

Sundog also liked that particular bed, not that he always used it. Usually he slept stretched out on the floor on Steve's side of our bed, but while he was waiting for us to retire, he liked to be in the middle of the room where he could watch us. This was why he wanted that special dog-bed. Unlike Ruby, though, he was too group-oriented to go upstairs before we did merely to occupy it. What's more, he didn't think he needed to—if he happened to find Misty or Pearl or even a cat on his bed when he approached it, he would stand right over her, staring straight down at her, silently demanding that she move. Misty and Pearl would willingly obey him. (So would the cat, or she would while Sundog was in his prime. After he grew old, she wouldn't budge. In his later years, if a cat were impeding Sundog, I'd have to move the cat for him. But that's another story.) Ruby would also obey Sundog, but usually not for a long time, and then with reluctance. Unlike the cats, she took notice of him, though. She'd hunker down, fold her ears, and avert her face, telling Sundog that she acknowledged his superior status, but that she didn't want to get up.

If Sundog persisted, and he often did, Ruby would get up eventually and slink away on bent legs, showing both humility and displeasure. Her attitude seemed to make a difference to Sundog. Perhaps he was afraid that she might successfully challenge his authority—after all, if she didn't move, there wouldn't be much he could do about it, as he rarely punished female dogs. Or perhaps there was some other reason as yet unknown. But whatever the reason, Sundog didn't always make Ruby move, so sometimes she got to stay on the bed.

Hence she always tried to. This meant that she was already upstairs when we called the dogs. At first, we didn't catch on, and kept calling. Any of the other dogs would have responded immediately, and would have come downstairs again just to oblige us, but Ruby knew that we were calling her in order to go upstairs, and thus she felt that she didn't need to obey us—she *was* upstairs, and hence had already obeyed, as she saw it. That she felt the mistake was ours, not hers, would become clear as we entered the room. She would look up at us and pat the floor with her tail, smiling pleasantly, with no shred of guilt, no suggestion that she thought herself in the wrong.

Nor was she, really. Autonomy is a wonderful thing, no less to an animal than to a person, witness the white-footed deer mouse of the daylight-darkness experiment mentioned earlier, the experiment which accidentally demonstrated that even a white-footed deer mouse wants to control his own life.

And Ruby was good at controlling her life, which evidently was an unseen by-product of being raised by indifferent owners and trained by brutal methods. Without leadership, without knowing what our species wanted of her, she figured things out for herself. She had never seen a dog-door before she came to live with us, for instance, and she stopped, bewildered, in front of it, but I had only to move the flap aside for her to get the picture, and after that she ran in and out at will, preferring the dog-doors to the human doors just as Misty had done. But then, unlike all our other dogs, including Sundog,

Ruby figured out how to open the dog-doors even when we wanted the dogs to stay in, when we put wooden covers over the doors and barred them. The bars rested in L-shaped hooks in the style of certain barn doors, and Ruby soon realized that if you put your nose under a bar and flipped upward, the bar would drop and the wooden cover would come loose. Then if you poked it a few times until it fell out of its frame, you could run outside to do whatever pleased you. Not a problem.

Fascinated by this interesting creature, I gave her the towel-biscuit test. She watched with much interest as I placed the towel over the biscuit. Evidently she didn't see the exercise as a puzzle needing to be solved, as had Sundog. Nor did she feel intimidated by the unfamiliar procedure, as had Misty. And she didn't take the towel to be a keep-off signal, as had Pearl. No—to my astonishment, Ruby simply grabbed the lump that was the hidden biscuit and crunched it right through the towel. Moments later she danced from the room having swallowed the entire test—biscuit, cloth, and all—leaving me speechless with surprise and my best kitchen towel reduced to a rag with a hole in it. Ruby was unconventional, it's true, but she was a dear little, smart little dog.

13

AS TIME went by, our household became more complicated. A neighbor rescued two stray female cats, who happened to be pregnant. The kittens were born, and the neighbor needed to find homes for them. I visited her with the intention of taking one, and came home with three, a brother and sister, both brown tabbies, and their half-sister, fluffy and black. We named them Jeoffrey, Kochka, and Machka.

Shortly before that time, my mother had left her home in Cambridge and had come to live with us in the apartment we had added to the house. She was in her mid-nineties, so we found four women to be her companions. The woman in charge was named Susan, and she would stay in our house if I was traveling. Eventually she began to bring her two dogs to

work with her. One of them, a neutered male, was a big, black, Irish setter cross named Hobbs. The other, Betty, looked like a sturdy brown beagle, and reminded me of our lost Fatima.

In addition, a neighbor named Don would bring his dog for us to care for in the daytime. Don was a divorced construction worker who lived in a log cabin on a nearby mountain with his power tools, his guns, his ammo, his canoe and power-boat, his ice shack and fishing tackle, his live bait, his mounted deer head, his mounted brown trout and lake trout, and also the aforementioned dog, who, contrary to what anyone might expect, was not a German shepherd or black Lab or rottweiler, but a little white fuzz-ball of a Shih Tzu named Wicket. Wicket became disconsolate if left alone in the cabin and he couldn't go to job sites to be confined in Don's pickup or to run around under the wheels of the heavy equipment, so early every morning, Don dropped him off with us. Don and Susan were friendly, he and I shared a garden, and on most nights he had dinner with us, which, thanks to him, was often trout or venison. If he wasn't a relative, he might as well have been, as might Susan.

For Thanksgiving that year, Don brought a turkey and cooked it in a smoker in our garage. It smelled so good that wild animals came out of the woods and yet another stray cat came up the driveway. The cat was a young male gold-colored tabby that Don named Krugerrand, I named Pula, and one of the guests named Chunky, and when the guests left that evening still talking about how delicious the turkey was, the cat stayed. Our household thus had five people—me, my husband, my mother, Don, and Susan—who were more or less permanent members, three other women who cared for my mother who were interim members, seven dogs, nine cats, and five parrots.

At dinnertime one evening, I looked around the kitchen as I prepared a variety of meals designed to please all members of our group, many of whom were sitting around me on the floor, watching my progress and waiting expectantly, and I began to wonder what had brought us to this point. We certainly acted like a family, but we didn't look like one. On the contrary, only the dogs and the cats resembled one another even remotely, having shared an ancestor back in the Eocene. Still, those 50 million years seemed short compared to the gulf that separated the dogs and cats from us, the people, as we had diverged during the Cretaceous, 80 million years earlier. Thus we were biologically closer to the mice in the walls than we were to our own cats, who obligingly hunted them for us. Yet all this paled beside the parrots, from whose lineage the dogs, cats, and ourselves had split during the Triassic or even the Permian, almost 300 million years ago, when a hefty, lizard-like creature who walked with his elbows and knees akimbo gave rise to the mammals and also to the dinosaurs and thus to the birds. Hence the parrots were much more closely related to *Tyrannosaurus rex* than they were to me. Yet it was me they called Mama. Was this natural?

It didn't always seem so, not if I looked at our kitchen through the lens of the Kalahari as it was when I first visited that pristine wilderness in the 1950s. Interestingly enough, all the creatures that were present in my kitchen were also present in the Kalahari, or close enough. People were there, the Bushmen. Dogs were there too, in the form of jackals—dogs and jackals being very closely related. Cats were also there—the little African wildcats who not only look like domestic cats but also belong to the same species. And parrots were there. African grey parrots exactly like our Pilgrim lived around the edges of the Kalahari, and Rüppell's parrots and peach-faced lovebirds flew in and out of the bushes that grew on the savannah, precisely the same species that are found in the homes of parrot fanciers today.

Yet the Kalahari was nothing like my kitchen. In the Kalahari, the people preyed on the jackals who preyed on the cats who preyed on the birds. The people sometimes preyed on the cats and birds too. Nowhere were the interspecies relationships anything that a pet owner might call positive. How was it that in New Hampshire, dogs, cats, birds, and people could gather together in a kitchen preparing to eat meals which did not consist of one another? Indeed, such a thought never crossed anyone's mind, except perhaps mine, and only then to marvel. What was going on?

Just at that moment, Pearl barked. She wanted her food, but all these ideas were distracting me, and I was just standing there in front of the kitchen counter, looking at the dog food, but not putting it into the bowls. In other words, I was being slow. Still, Pearl managed to communicate her wishes. *Ah yes. The dog food.* I returned to the task at hand.

I hadn't finished filling the bowls before I had to stir the human food in a wok on the stove. Pearl barked again, staring at me. I wouldn't be able to serve her meal for a minute or two, so I handed her a dog biscuit. She took it, but held it in her teeth and continued to stare. My progress was too slow for her liking. Then she barked once more, just as if her mouth were empty, which was her special skill. "Please don't bark at me, Pearl," I asked her.

The food in the wok was smoking. I turned back to the stove. As I did, Pearl barked again, this time more urgently and sharply. A cat was on the kitchen counter and, taking advantage of my inattention, was helping herself to what little food I had so far managed to put in the dog bowls. The cat's dinner, too, was late. I turned around in time to see the cat leap across the sink to the far side of the counter as Pearl, now the very picture of tension, stared at her fiercely. Pearl didn't want the cat to take her food, and the cat knew it. I heard my own voice say, "Pearl," rather contentiously. Unaware that I had spoken, I looked around. There stood Pilgrim, the African grey parrot, on the back of a chair in the corner. He had a way of anticipating what people would say, and saying it for them. But Pearl disliked being spoken to by parrots, especially not in my voice. She turned away and sat down stiffly, her ears partly folded, facing the opposite wall so she didn't have to look at Pilgrim. Perhaps her motive resembled a person's, or perhaps not, but she deliberately turned her back to him. Thus the members of our household dealt with one another. We had come a long way from the savannah.

How did this happen? Dogs made it happen. Their ancestors, not ours, are responsible for animal domestication. That we have dogs to thank for the long period of time we have spent in their company can be said with some certainty, and is not really a matter of conjecture, although in the past we have taken full credit. We sometimes hear stories about how early people went out, found a wolf cub, brought him home and tamed him, but the process could not possibly have been like that. For one thing, wolves are just as social as we are, but don't bond with people as well as dogs do. Human companionship would not have been enough for the pup, and as soon as he grew up and heard his relatives singing in the forest, he would have gone to join them. For another thing, just as modern hunter-gatherers and also Third World villagers don't have surplus food, our Paleolithic ancestors didn't have surplus food either, and like their modern counterparts, could not have wasted what they had on animals. In the camps of modern hunter-gatherers, and also in Third World villages, most dogs are very seldom fed, and some are never fed at all. Feeding dogs is a new idea, relatively speaking. The wolves who first approached us came as scavengers.

Furthermore, the process that led to animal domestication still goes on right under our noses, where we can see it. Bats, sparrows, mice, rats, raccoons, squirrels, and other animals invade our dwellings hoping to find nesting places, to share our food, and to find protection from predators. Some of these animals have the necessary qualities for domestication—rats are social, for example; also they accept leadership and they eat the same foods we do, but we haven't yet capitalized on these qualities in rats or found places for them as domestic animals. But that's our fault, not theirs. They are taking the first step.

As for wolves, they lived in Europe and Asia for a million years before our ancestors came to those cold places, and surely they noted similarities between themselves and the newcomers. Like them, the people lived in groups, and needed to protect their young from harsh weather, to which end the wolves dug dens and the people built huts or sought out caves. Both species hunted in groups for animals larger than themselves, depending on cooperation to bring these animals down. Wolves and people also fed their young in much the same manner. Many animals bring their young with them to a food source, eat there, and travel on. Wolves and people do it differently, as only the adults forage, leaving their young behind in shelters and bringing food back to them, the wolves carrying the food in their stomachs and the people carrying it in bags. No two animals could have been more similar than these carnivores and these primates, something that the wolves must have seen.

The wolves would have visited the people's habitations just as, in certain areas, wolves still do today. A human encampment produces edible scraps which a wolf could find when the people were elsewhere or sleeping. Here again is evidence suggesting that wolves approached us, not we them, as we would not have been visiting their campsites. There isn't much to see at a wolf den, where the activity is mostly underground, and food arrives in the form of vomit, which is instantly eaten by the pups. A human encampment, in contrast, is often a hive of activity where edible scraps abound. Bears and other predators would also know about the scraps, and when they came to investigate them, the wolves would vocalize to notify one another of the problem, and thus would notify the people inadvertently.

What's more, unlike some of the other predators, the wolves probably did not hunt people. The reason for this is unknown, but in all recorded history, there is not one verified case of a wolf making an unprovoked attack on any person anywhere, and much research has been done on the subject. However, other predators were dangerous, so the warnings given by wolves would have been useful. Possibly the wolves would sometimes drive the dangerous animals away from the human encampments, as they would from their own dens.

More scraps, or at least bloody snow, would be left behind by the people in the aftermath of their successful hunts, and this food may have induced the wolves to drive game for people as they did for one another, in hope of more scraps and more bloody snow. It would not have taken the people very long to see what the wolves had already seen—the benefits of a human-wolf alliance—and the future of the dog would have been assured.

I was twice fortunate enough to witness situations that, had they taken place in earlier times, could have encouraged animal domestication. In the 1970s, I participated in a study of wolves on Baffin Island, where I lived alone near a wolf den. At the time of my visit, the interior of Baffin Island was relatively unexplored—the maps were inaccurate, anyway—and no people, including the Inuit, had ever lived there. Hence, as in Paleolithic times, the wolves were not afraid of people. They were cautious, but also curious, and would visit my camp to learn what they could about me. They took my only sweater, for example, and not knowing what it was, they shredded it to learn more about it, forcing me to make myself a garment from a towel. I didn't have much food, and what I had was under a pile of rocks, so they didn't get much to eat from my camp. If they had, they would have come there more often. As it was, they occasionally followed me at a distance when I explored their trails just as wolves would have followed the Paleolithic hunters, probably because they wondered what the bizarre creature, my bipedal, primate self, was doing.

Of course, I welcomed their presence. Not only was I trying to observe their denning practices—my reason for being on

Baffin—but also I had been told of a polar bear who sometimes visited the area. That worried me. People are about the size and shape of seals, and understandably fall prey to polar bears. I had no weapon or pepper spray of any kind, and because I slept squeezed into a little cave just barely big enough for my body, I was afraid that the bear would find me and dig me out. But the wolves didn't want the bear around any more than I did. If he had come, they would have warned one another and I would have heard the warning. Thus they would have helped me in the same way that, long ago, the ancestors of our dogs helped our Paleolithic ancestors in their camps.

Earlier, in the 1950s, I saw an equally preliminary aspect of the domestication process when I lived in the Kalahari among the Bushmen. Predators large and small would occasionally visit their camps. (Not surprisingly, only the predators visited, and other animals did not. Members of the antelope family, for instance, came nowhere near the Bushmen.) Lions sometimes came, evidently for no better reason than to view the people at their fires, but of course the people didn't want them there and would tell them to leave in no uncertain terms.* The lions might then rumble a bit, but they'd go. Hyenas also came, especially after the people had killed one of the large antelopes and had meat in their camp. Once I noticed a hyena taking away one of my shoes. When I asked her to give it back, she gave me a look, and dropped it.

On rare occasions a leopard would come, probably with the more sinister object of hunting people. Since the Bushmen slept lightly and didn't all sleep at the same time, someone always seemed to notice the leopard, and drove it off. And jackals came, to poke their delicate noses through the grass around the campsite, hoping for the tiniest of scraps. To me, the jackals were among the most interesting animals because they are

*The relationship of lions and Bushmen is discussed at length in Elizabeth Marshall Thomas, *The Tribe of Tiger,* New York, Simon & Schuster: 1994, 127–86, and "The Old Way," *The New Yorker,* October 15, 1990.

very closely related to dogs and wolves and their behavior is similar.

The jackals may have had another reason for lingering near the Bushman encampments. Because jackals are occasionally the prey of larger animals, especially leopards, they may also have found some slight protection in the presence of the Bushmen. By keeping the large predators away from the camp, the Bushmen would inadvertently have kept them away from any jackal who happened to be nearby. Like the Paleolithic peoples long ago, the Bushmen were not concerned about these social members of the dog family, and because the jackals remained inconspicuous, they were more or less ignored. In the morning, one could sometimes find the tracks and also the scats of the nocturnal visitors. Fossilized scats of wolves have been found by archaeologists in some of the Paleolithic campsites. As the Bushmen of the 1950s more or less ignored the jackals, so the Paleolithic peoples probably ignored the wolves.

Occasionally someone asks why the Bushmen didn't domesticate jackals. They certainly had enough time to do so—archaeologists found evidence indicating that people have lived in Bushmanland for 35,000 years at the very least, with a material culture very similar to that of the modern Bushmen. But that's the wrong question. The right question would be: Why didn't the jackals take the necessary steps? And probably there are several answers to that. First, jackals aren't quite as social as wolves, which is to say that their groups are not as large and, more importantly, not as catholic—under the right conditions wolves will accept unrelated wolves as group members (hence one doesn't always need to be a relative to belong to a wolf pack), but as far as I know, jackals are less likely to accept unrelated jackals. The scraps they find around a campsite seem to be all they want from our species, if that. Why would they, in their close little units, decide to pursue matters further?

In addition, jackals lack one of the most important incentives that drew wolves to the Paleolithic camps. The wolves and the Paleolithic people hunted the same prey, but the jackals and the Bushmen did not. The jackals hunted small animals and the Bushmen hunted large animals. To be sure, the Bushmen certainly took the odd porcupine or springhare and thus occasionally competed with jackals, but it was the game antelope that the Bushmen strongly preferred, and hunted most often, by a very substantial factor. To be sure, both the Bushmen and the jackals hunted the smaller antelope such as springbok, but again without competition, as Bushmen hunted the adults of these species, and the jackals hunted the fawns. A fawn or a porcupine does not provide enough meat to encourage cooperative hunting—it's too small to share.

In the Kalahari, it was lions, not jackals, who resembled the wolves of the Paleolithic campsites. Like wolves, and also like ourselves, lions are a social species. Also the lions and the Bushmen hunted the same game in much the same manner, camped in the same kinds of places, and depended on the same sources of water. Until the 1970s when the Bushmen acquired cattle, the lions and the Bushmen had a truce, so that neither group interfered with the other. If an alliance could have been formed, that would have been the window of opportunity. But while the Bushmanland lions have not hunted people in historic times, they might have done so earlier, may very well do so again, and certainly do so elsewhere. The Bushmen might have ignored any jackals who came around their camps, but they certainly didn't ignore the lions for the very same reasons that make lions poor candidates to assume a doglike role.

In contrast, wolves and their descendants have reached a climax of domestication although, as is true of the first stage, we in the United States must travel elsewhere to see it. Europeans are vastly more civilized than Americans when it comes to dogs—too many Americans are veritable dog fascists whose need to control is so overwhelming that they would have all dogs caged if they could, but in other countries, dogs are appreciated in public parks and are allowed on public transportation where they calmly ride on subways and on escalators. Dogs are also allowed in many restaurants, where they are served bowls of water and leftover food. Some restaurants even have menus for dogs. In all cases, the dogs must be well behaved, of course, but because dogs are so adaptable, so aware of what is wanted of them, their public behavior is seldom a problem. In Prague, for example, it is not uncommon to see a dog all by himself on a crowded sidewalk, trotting along among the people with a folded leash in his mouth. Why the leash? Czech law requires dogs to have leashes, but doesn't specify where the leash must be. The dog's walk is purposeful—he knows exactly where he is and where he's going, and his owner is probably around somewhere too. The dog isn't lost and he isn't causing trouble, and nobody objects. On the contrary, the few people who take notice of the dog seem to approve of him.

The most engrossing scene I witnessed was in Milan, in the lobby of a hotel. I happened to look out at the street through the hotel's automatic glass doors and noticed a golden retriever on the sidewalk. He approached the doors and they slid open for him. As he walked in, the doors closed again behind him. No one in the lobby seemed aware of the dog, so I was a bit alarmed, as I thought his owners must be out on the street and wouldn't know where to search for him. If in the United States a dog had walked into a hotel lobby, people would be frantic, running around, waving their arms, yelling at the dog, yelling at one another, and calling 911. Expecting to hear sirens and see flashing blue lights at any moment, I looked at the doorman to see what he made of this, but he seemed not to have noticed the dog

The dog was unworried. He calmly sat down in the entrance area and glanced around briefly. At first he seemed to see nothing unusual, and then he spotted me watching him. At once, his ears lowered and his face softened, and he gave me a quick dog smile of acknowledgment just as a person might nod a polite hello. Then his gaze returned to the street.

Time passed. As I was still the only person who seemed to have noticed the dog, my concern was slowly mounting, and I was about to point him out to the people who were with me, when the automatic doors opened again and two children, about eight and nine years old, also came in. The dog gave them the same kindly glance that he had given me, this time flapping his tail once or twice, but he didn't get up. The dog and the children then waited together. More time passed. Eventually, the automatic doors opened a third time to admit a beautiful woman, surely the children's mother, who swept right past the dog and the children and headed straight for the bar. The dog stood up and he and the two children calmly followed her.

Obviously, the dog knew exactly what the woman was doing. He showed no anxiety or eagerness, just pleasant anticipation while she ordered an aperitif and a plate of sandwiches—he knew he'd get his share—and he stayed politely by the woman's bar stool while she chatted with the bartender. When the sandwiches came, she passed them to the children, then delicately held one out where the dog could see it. He stood on his hind legs, put his front paws lightly on the bar, and carefully took the sandwich in his mouth. In this genteel manner, without haste or greed, the family shared the sandwiches equally until all were gone.

The woman then slid down from her stool and swept off toward the elevator, followed by the children and dog, who politely waited for all the people, including his family, to get in first, then stepping neatly inside just as the doors were shutting. At the very last moment he quickly tucked his hips to get his tail in too. No person could have managed more smoothly. Serene, competent, pleasant, and urbane, this dog was like an upper-class Italian.

He was also a subspecies of wolf, unlikely as it seemed, no matter how far the old association between his kind and ours had brought him from the hunting camps at the edges of the glaciers—from the Arctic storms, the firelight, the reindeer meat, the bloody snow. The old ways are still very much with dogs, for although we have done much to change their appearance we have done nothing to modify their wolfish hearts. It is this very nature that has made them so adaptable. No sheep or horse would know how to act in a four-star Milanese hotel.

I once had the pleasure of visiting the singer Wynonna Judd at her home near Nashville where she sheltered dozens of animals, mostly dogs and cats, mostly rescued strays. Among her many pets were six or seven very small dogs, all toy breeds, including (if memory serves) a miniature poodle, miniature Pomeranians, Chihuahuas, and the like. These fetching creatures certainly didn't look like wolves, but it was fascinating to see how much they acted like them—the tiny things had worked out a wolflike hierarchy with an alpha member who led the rest in a foraging trip through Wynonna's living room, speeding across her white carpet as if across a snowblown tundra, climbing on her white upholstered chairs as if up mountain ice fields, leaping over the cracks between her white sofa cushions as if avoiding crevasses, and pushing through the folds of her gauzy, floor-length, off-white curtains as if through frosty thickets on the banks of an Arctic stream. Standing in the middle of the room and looking down on the display of wolfishness that swirled around my ankles, I might have been none other than the great wildlife biologist L. David Mech himself, circling Elsemere Island in an observation plane. In Wynonna Judd's white living room, the spirits of the wolves who joined our ancestors were running as a pack.

14

WYNONNA'S dogs were a cohesive group, the Italian dog and his Milanese family were too, and so were the wolves who approached the Paleolithic camps 100,000 years ago. And so were we, in our multiple-species household in New Hampshire. Nothing is more important to a dog than group membership. Some people believe that sex, food, and personal safety are the drives that motivate an animal, but in the case of dogs, they're wrong. All these evaporate in the face of group requirements, so that a dog will not eat if higher-ranking dogs coerce him, or if his owner doesn't want him to, and sled-dog drivers can put bitches in heat on their dog teams, knowing that the bitches will not be bred, just as low-ranking wolves in heat will sit down if the high-ranking fe-

male gives them a stare when the males come around. The group is important. The individual is less so. The individual subordinates himself to the welfare of the group. Thus, like a finger of a hand, a dog by itself is almost not a thing. Only as part of the hand does it have function.

One day I came home after an absence of several weeks. My husband and I had gone away together, during which time my mother's companion, the wonderful Susan, had stayed at the house, bringing her dogs with her for the first time. I came home ahead of my husband, and was met at the door by Susan's dogs, who had never seen me before and quite naturally began barking. But my own dogs stormed past them to greet me, jumping and yelling as though they had feared that their group had split forever and we would never meet again. For just a fraction of a second, Susan's two dogs seemed baffled. But almost instantly they too began to greet me as wildly and enthusiastically as if they had known me all their lives.

Of course, this wasn't the case. Perhaps I didn't seem a total stranger, as surely they had found whiffs of my scent around the house, but this does not explain their welcome, as familiarity alone does not guarantee such enthusiasm. No, Susan's dogs had taken their cue from my dogs, with whom, after several weeks together, they had come to feel a sense of group. Evidently, the behavior of my dogs showed Susan's dogs that I was every inch a group member, and without knowing anything else about me other than my strangely familiar scent, Susan's dogs welcomed me as such.

Our four dogs had their own views about group membership, views which said much about them. Ruby barked at no one, at least, not during the first few years she lived with us. Instead, she was unfailingly friendly to all visitors from the moment they came to the moment they left. At the time, she didn't feel that she was fully a member of our group, and wanted to join someone else.

Misty was so focused on keeping Place Two that she barked at anyone, everyone, as in her eyes, anyone might take Place Two away from her. Sweet Misty, she sometimes even barked at me because she saw danger in every arrival. Thus to her, I was a threat first, and an individual second. Then she would feel apologetic. *Oh dear, I barked at Liz,* her manner said, and she would carefully search my face with her worried, black eyes. Of course, I always told her it was perfectly okay, that she'd done nothing wrong, that she was a dear, good dog, and we loved her. But after making such a mistake, she seemed not to believe me.

As for Pearl, Pearl wanted to bark at everyone, but she spared Steve and me. She also spared our son and his family, the people who had raised her. They lived in her heart, and although they came to visit only once a year, she would greet them wildly, joyously, just as she greeted us after an absence. She has been with us for ten years at the time of this writing, and once a year, every year, she continues to recognize and welcome them.

She would, however, get news of them more often than she saw them, which may have helped her to keep alive their memory, because I visited their house every so often, where I would wade through the cloud of dog-detectable odors that lay low in the rooms. When I came home again, Pearl would treat these odors as if she were a person reading a letter. She would vacuum my clothing with her nose, starting at my ankles and working her way up my legs to my knees. On my boots and in the cloth of my jeans she found her father, also the other dogs who had raised her, also Manas the cat who was shaped like a beanbag, also the rabbit who hunted squirrels, also my son, his wife, and our grandchildren, all beloved members of her first group, of which my husband and I were, to her, an extension. When Pearl had finished her investigation, when she had gathered all the news she could about these animals and people, she would sneeze, then look up at me with a quick, meaningful glance as dogs do with one another—just checking to see if both of us understood what it was that she'd found.

Do dogs feel emotion upon sensing distant group members? Pearl did, and the sneeze showed it. Dogs don't often sneeze as we do because of dust, pollen, and the like, and if they do, they normally sneeze heavily and repeatedly. But a single, deliberate sneeze is a kind of conclusion, usually after the dog has been having strong feelings, often after punishment or after prominent displays of affection or approval, but also after feelings such as those experienced by Pearl from the scent of her loved ones in Boulder.

Yet Sundog's sense of group was the most remarkable. Of course, he didn't bark at me and Steve, but with certain very interesting exceptions he barked at everyone else. The exceptions were my mother, our son, our daughter, their spouses, and our grandchildren. Unlike Pearl, however, Sundog had never lived with any of these people, and, until my mother came to live with us, he saw them only once a year and only for a few days at Christmas. He might bark when he heard their cars in the driveway, but the moment the car doors opened and he saw who was inside, he would push past anyone standing in his way to greet these people gently and warmly, his chin high and his ears low.

Equally surprising, perhaps, was his soft manner when offering his greeting. This was quite unlike the wild feelings he displayed when he saw me or my husband after a long absence. Thus he saw a sameness between us and our immediate kindred, but he also saw a difference, as his different greetings showed.

I can only explain certain parts of this miracle. That Sundog remembered these people is understandable, since many animals are known to have excellent memories, a fact that has been demonstrated scientifically again and again. As for our own animals, I'd say that they remember people well if the relationship was important. We acquired one of our cats from a family that had too many cats, mainly because the woman of the family didn't believe in having cats neutered. During a time of personal stress, the woman decided to get rid of all her cats, and bundling them into boxes, she took them to the humane society, where she put them up for adoption. In fact, since many more cats are offered for adoption than are taken, these cats were probably euthanized.

One of the cats, a cat whom the woman had always disliked, realized that something dreadful was happening to the other cats and escaped the purge by hiding in the basement. She eluded the woman from that time on by hiding in closets and under furniture. Learning of this, I offered to adopt the cat, and her owner gave her to me. The cat lived happily with me for many years. One day, however, the woman in question came for a visit. She entered through a door which was right beside the kitchen counter, where the cat happened to be sitting, watching me prepare her food. The cat looked up to see who came in. There stood her former owner. The cat hadn't seen the woman since the time of the purge, but she certainly remembered, and the sight gave her a terrible shock. Her hair rose, she crouched, flattened her ears, and spat like a little thunderclap. Then she flew off the counter and sped away, not to reappear until the woman was gone.

So an excellent memory might explain how Sundog recognized our children. It does not, however, explain how he knew that I, my mother, my husband and our children and grandchildren were somehow a group. This seemed especially puzzling in that he did not feel the same about Steve's brother, or about my brother, both of whom he saw more frequently, but barked at. Nor does Sundog's good memory explain why he greeted Steve and me wildly, but greeted our offspring more calmly. As for the latter question, if I were to guess, I'd say that

Sundog needed me and my husband, and perhaps felt desperate while we were gone, hence he greeted us wildly to express his joy and relief. He didn't need my mother or our children and grandchildren in the same way. Nevertheless, they were his own, just as Steve and I were his own, more so by far than all the other people whom he saw every day.

His attitude toward our grandchildren was particularly striking. He didn't like young human beings, yet when the first of our grandchildren was brought to our house as an infant only a few days old, he sat perfectly still and stared at her in awed silence for almost half an hour, and never failed to recognize her instantly thereafter, although between visits she grew up bit by bit, so—unlike an adult, say—her persona would have been different every time.

Sometimes, a dog's sense of smell is credited for such feats. This was not the case with Sundog, as his recognition of our grandson shows. The grandson was adopted by our son, and thus is not biologically related to us and would not carry any genetically determinable odor, assuming that there is such a thing. Also, being considerably older than his sisters, the grandson sometimes came to visit us from his college dormitory, and would carry no familiar household scent. Finally, our grandson sometimes came to visit on his own, without his parents, and thus, as far as Sundog was concerned, was out of context. Even so, Sundog unfailingly recognized him each time he came, and didn't bark. All in all, it seemed uncanny.

Since organized data speaks more powerfully than random impressions, I once tried to assess the number of times per year that Sundog had to deal with a person arriving at the house, and thus how often he needed to determine whether or not to bark. I estimated at least 1,000 arrivals, some by people who came every day and went in and out several times a day, some by people who came every weekend, some by people who often visited, some by people who seldom visited, and some by people who were delivering packages, bringing the mail, or reading the electric meter. The count did not include me or my husband. Otherwise, of the 1,000 arrivals during that year, only eight, just eight, were by our children or grandchildren, or less than one-tenth of one percent. Yet Sundog was never wrong. All on his own he would flatten his fur, soften his face, and greet his people with much affection. *Ah, my dears, you've come at last,* his manner said.

The questions that cannot be answered are as interesting as those that can. One of these was posed by Pearl, who made only one exception to her rules of barking. This was when a car arrived bearing Jeffrey Moussaieff Masson and his wife and two-year-old son. It was their first and only visit to our household, and none of us, least of all Pearl, had ever met them before. Even so, she did not bark. She stood there looking at them, and did not bark. I was so astonished that I was probably impolite to my guests, and I tried hard not to bore them by talking too much about this miracle. True, Jeff Masson knew animals well, and cared about them. He made this abundantly clear in his writing. But then, my mother also liked animals, and Pearl barked at her.

I immediately assumed that Jeff's presence had inhibited Pearl's barking. Here he was, getting out of a car, the famous writer whose fascinating work meant so much to so many people. If I were a dog, I certainly wouldn't have barked at him. More likely, I would have greeted him wildly. But he wasn't the only person in the car, so perhaps Pearl's reason was different. Perhaps the person she decided not to bark at was his wife, or, even more likely, his little son. Pearl didn't often meet a human baby, but she knew what she was seeing. As the family got out of the car, Pearl stood quietly at my side and watched them tenderly.

15

NEVER did our dogs more clearly display their sense of group than in their response to animals of Category I, other dogs. Their noisy response to visiting dogs made our house almost uninhabitable until they all calmed down. Touchingly, an echo from the woods was also Category I—it too was a dog. Each of our dogs discovered this echo independently shortly after he or she had joined our household, noticing the voice in the woods while barking at something else. Perhaps the dog would have seen a deer, and would have gone partway down into the field to drive it off, when suddenly a big dog in the woods would start barking back.

Forgetting all about the deer, our dog would bristle and bark furiously. The dog in the woods would of course respond

in kind—his challenging voice ringing out whenever our dog paused to listen. Our dog would bark at him again. Perhaps dogs don't recognize the sound of their own voices, just as we don't recognize ours on a tape recording. When Pearl first discovered the echo, she barked passionately for forty-five minutes without pausing, suggesting that she had no idea that it was her own voice she was hearing, and also, incidentally, that she might be able to bark while breathing in as well as while breathing out.

She was standing when she first engaged the challenger, of course, the hair on her back aloft. But in time she grew hoarse and also tired, and she sat down to listen, still facing the dog in the woods. He made no sound but he was there, Pearl knew it, and his silence must have seemed as menacing as his voice. Pearl stood up again. He didn't seem to change his tone in response to her challenges, so who knew what he was thinking, or worse, what he was doing? Nervous, she resumed her barking, perhaps so that he'd respond and she'd know where he was. He answered in exactly the same way from exactly the same place. Had he taken over the woods?

On and on they dueled, their voices ringing, until evening came and Pearl gave up and went into the house. But not for long. The thought of the other dog troubled her, and after eating her dinner she went out to challenge him again. He was still in the woods, just where she'd heard him the first time. She barked until we forced her to come in.

Interestingly, all four of our dogs, all on their own, eventually figured out that something was strange about the echo and ceased barking at it. But the thought of the woods-dog continued to unnerve them. Sometimes they would catch the echo as the result of unrelated barking, and would stop, sit down, and face it. They didn't bristle or bark, though—instead, they seemed to be pondering the problem. I was reminded of the occasions when they first saw themselves in a mirror. Many dogs behave in more or less the same way when they first catch sight of themselves—they see the reflection and stop dead, ears forward and hair rising. *Whoa! A dog!* they seem to say. They may even bark, but not lengthily—just a single, muted *woof* of surprise at the sight of that other dog staring right at them. But almost immediately, they realize that the image isn't what they first supposed. They may not even bother to sniff at the image and they never seem to investigate it further. Instead, as if embarrassed, they assume a humble stance and back away. They also avoid the mirror thereafter. If dragged in front of it (I sometimes did this for an experiment) they look at the reflection blankly, as if the image in the mirror was no more important than the wall behind it. It is often said that dogs don't react to their reflections in a mirror. Ah, but they do. I submit that if a dog does not react, he has already seen his reflection earlier.

Hence the reaction to the echo was something like the reaction to a mirror. The dogs just took longer to catch on, probably because if a real dog were in the woods, he would be hidden by trees and also at a distance. Hence the echo was more difficult to spot as a counterfeit, whereas the dog in the mirror was right there, inches away, and had no odor. But even after they stopped barking at the echo, doubts lingered. As each new dog discovered the echo, the experienced dogs would join the new dog in the field, and while the new dog barked, the experienced dogs would all stand around nearby, facing the woods uneasily, not joining the new dog's continuous frantic barking but woofing firmly every now and then and remaining alert for developments. They knew the echo wasn't right, but here was the new dog, dueling with it. That alone must have seemed compelling. Such is the power of the group.

Yet it was coyotes that evoked the most primal reaction in our dogs, as if, when faced with their own wild relatives, their long sojourn in our company meant nothing, and they returned to a wolfish state. To our dogs, nothing was worse than the sight of a coyote minding its own business, hunting mice in the field. When the dogs spotted such an intruder, they would bark, but only very briefly—not with loud paeans of barking such as dogs use when they must inform the dense, uncomprehending human species—just a muffled yap or two to get the attention of the other dogs. In this same manner, wolves inform one another.

Walking stiffly, filled with purpose, the dogs would assemble at the edge of the lawn and watch the coyote briefly, then, all together, they would run into the field, gaining speed as they approached him. But about fifty feet away from him, they would slow down, spread out, and then more or less stand still, showing him their sides.

To me, this all seemed meaningful. That the dogs went to the field together in a pack was in itself important. When dogs are ranging around for no special reason they usually travel in pairs, and if a third dog is with them, it strings along far behind. Wolves also travel singly or in pairs, or they do in denning season, while the den is their locus, where eventually they'll all meet again. By going out at different times and in different directions, they can take turns baby-sitting the pups, and they can also maximize their chances of finding game. We think of wolves as traveling in packs, and so they do, but only after the denning season is over, when they range far and wide over vast distances and have no special place to reassemble. Thus they must stay together. Hence dogs are like wolves during denning season. Why? Because the house where dogs live is their locus. They stay there year round, and they always know where to reassemble. They don't need to form packs except under special circumstances. The most compelling of these is to confront an invader.

Therefore that our dogs would run out as a pack to meet the coyote showed what they thought of him—he was an important invader. But why, in that case, didn't they then continue their charge? Why did they stop and stand around sideways, near him but not facing him?

Whatever the motive, all of them understood it, the dogs and the coyote too. But we, the human beings, didn't understand. Instead, we'd get worried and start calling the dogs to come back. However, by this time, our dogs would be listening to their wolfish inner voices, and none of the animals in the field, not even the coyote, would seem to hear us. The coyote could clearly see us if he chose to look in our direction—we would be standing at the edge of the lawn. If not for our dogs, he'd have run for the woods when he heard our voices. But faced by the dogs, he wouldn't react to us because he didn't want to seem doubtful. Like any animal in a tense situation, he would want to maintain a cool, self-confident facade.

As for the dogs, the encounter itself, with all its primal qualities, would engulf them. Dogs in tight spots often glance at their group members to learn if the others see what they see, or know what they know, and we were certainly their group members, but when our dogs faced a coyote, they would glance frequently at one another but they never glanced back at us to learn what we made of the situation. No—at this point in the encounter, they would have eyes only for one another, and also for the coyote, and he for them.

And there they would be, presenting themselves side-on, glancing at the coyote over their shoulders. To face him head-on would have been confrontational. To show him their sides was softer, more appeasing. But also they were showing him how big they were, and that he was outnumbered. They could, of course, have challenged him, facing him, showing him their teeth, barking aggressively, and staring—ordering him off the place, as it were. But if they did, he might have thought he had to fight them. Instead, they were waiting to see if their display would do the trick, so that the idea of leaving would come from him.

Usually the coyote would fool around a bit at first, as if he didn't care what the dogs thought, but sooner or later he would return to the woods in a slow, indifferent manner. Sundog, Pearl, and Misty might follow him at a distance to the edge of the woods, but then would come swaggering home with their hair still bristling, obviously still experiencing the encounter.

But Ruby would sometimes follow the coyote. This was before she felt at home with us, when she still felt the need of a group, when she made up to any newcomer, whether she knew him or not. I thought that she might want to join the coyote. But would the coyote want her company? The first time Ruby ran off like that I was afraid the coyote might be female, and would tear Ruby apart if she got her alone in the woods, so I called Ruby frantically. But in those days Ruby didn't want to be a good dog and she ignored me. I then ran after her, which merely drove the two of them farther off. But evidently Ruby knew what she was doing. She was gone for an hour or more while I called and bit my nails and worried, but she came back unharmed.

The dog-coyote relations reached their climax in a textbook demonstration of territorial claims. One day the dogs and I found four scats just where the field joined the lawn—doglike scats containing hair and bits of bone. Obviously, they had been left there by coyotes. This of course was very clear to the dogs too. When we came upon the first of the scats the dogs stopped dead in their tracks and bristled. *Whoa! Important!* their manner said. Forgetting all about me, not even giving me a glance, they then trotted busily back and forth along the edge of the lawn in determined silence with their noses to the earth, obviously to learn as much as possible about whoever had done this. Thus they found the three other scats. No coyote had ever done such a thing before, and in this display, more than one coyote had participated. *The field is ours, from the woods to the lawn,* the scats said. The dogs exchanged meaningful glances.

It was all very interesting. The coyotes had left their scats exactly where the field joined the lawn although there was no wall or fence or any outward sign except for the type of grass. And the field had been mowed, as had the lawn, so the grass was short on both sides of the border. But then, most dogs have no difficulty perceiving territorial borders—at my mother's house in Cambridge, during a border dispute between a neighbor's German shepherd and a pug who for a time lived with my mother, both dogs mysteriously knew the exact location of the unmarked but legally recorded property line between my mother's lot and the lot next door although the two lots shared a single lawn which was mowed and cared for by the same person. Thus, except for a short, very distant fence that divided the lots at the rear, there was no outward sign whatever to show the dogs where one lot ended and the other began. Even so, snapping, snarling, and shouting at each other, the two dogs would run up and down the invisible line as if an actual, physical barrier stood between them, each claiming his lot with the accuracy of a surveyor. It fairly made one's skin prickle.

The day after finding the coyote scats by the lawn, I found four dog scats at the far edge of the field where a trail goes into the woods. The dogs never went that far merely to relieve themselves, nor did they normally view elimination as a group activity. This time was different. They had gone the distance to reclaim the field. So their scats too contained a message. *Ours from the lawn to the woods,* they said.

16

BECAUSE our dogs would form an all-dog group only on rare occasions, I couldn't help but notice how different they were from the dogs we had known before, of whom the last were Suessi, Fatima, and Inookshook, the faltering, elderly dogs who had come with us from Virginia. That group of dogs had been constantly together. Compared to them, Sundog, Misty, Pearl, and Ruby could hardly be called a group at all. They stayed with people, and very firmly too. If I didn't see Steve during the day, I wouldn't see Sundog either, because he would be with Steve. In contrast, Pearl was always with me. Our former dogs hadn't felt like this. Why were Sundog, Misty, Pearl, and Ruby so different? Had they taken us as pack leaders?

To be sure, Suessi's parents had formed a pack in the usual manner—they were a mated pair with some of their offspring, a group such as wolves would form. Other, unrelated dogs who joined them later understood the arrangement, and fitted themselves into the appropriate niches as subordinates, again just as wolves might do. Sundog and Bean might have formed a pack if they could have continued their relationship. Misty probably would have formed a pack with Sundog if he had wanted her. But none of this happened, so our dogs had no choice. Even so, this does not explain why they stayed so closely with us, the people, in tightly bonded, one-on-one relationships—or why they didn't later bond with other dogs who joined us. Nevertheless, as any pet owner knows, their choosing people over other dogs was absolutely normal.

But why is this so? Why, when dogs can choose other dogs, do they choose us instead? There are several common answers to this question, one of which is that dogs are like young wolves, so they accept us as their leaders, as wolf pups would accept older wolves. When dogs lay back their ears and raise their chins to lick our faces, they're acting like wolf pups begging food from their elders. Also their bodies have youthful features, such as loose skin, lopped ears, blunt noses, soft hair, and the like. Neoteny, the phenomenon is called. Hence, when we ask why most dogs tend to associate with people in preference to one another, neoteny is usually the answer. Just as a young wolf if given a choice would follow his parents rather than stay behind with his litter mates, so our dogs choose us over other dogs. At least, that's what the literature tells us, and surely there's some truth in it.

However, that answer is far too simple, and doesn't take the past life of dogs into account. Modern dogs share certain characteristics with young wolves, surely, but that's far from the whole story. There may be a better answer, and if so, it does not necessarily come from wolf times, but from later times, after dogs began to associate with us.

Genetic evidence suggests that the wolves who became dogs have been with us for more than 100,000 years. That's a long time, and during most of it, dogs did not live as they do today in most of the United States. Their sojourn with us began in the hunting camps of the Paleolithic people, and after that, in Neolithic villages where the conditions, from a dog's point of view, resembled those of the Paleolithic camps and also of certain settlements in some of the less-developed areas of the world such as parts of Africa and Asia. In other words, a dog is more than just a young wolf—a dog is a young wolf who has spent 100,000 years in Paleolithic camps and Third World villages.

And that wasn't easy. In the 1980s I did some fieldwork in Namibia, and while there I revisited the Bushmen I had known in the 1950s. By then, they had been forced to give up their hunter-gatherer lifestyle, and had acquired a small herd of cattle, and also six dogs—not the descendants of the little jackals that used to haunt their camps, but wolf-derived pie dogs or pariah dogs obtained from their Tswana neighbors.

In keeping with the practices of most Third World villagers, the Bushmen very seldom fed these dogs. Other Third World villagers don't feed their dogs either. As has been said, feeding dogs is a recent innovation. This may seem harsh to us in the United States, where obtaining food is not a problem, but in much of Africa and also in other Third World areas, food is always in short supply and cannot be shared with animals. The people give it to their children instead, a practice that has kept our species on the planet. Then too, in Bushmanland, the people feel that all animals should take responsibility for themselves, as wild animals do. So in Bushmanland, the village dogs scavenged, finding what scraps they could and eating human feces. Hence these dogs were very much like the earliest dogs. Almost certainly, the members of their lineage had never been fed regularly by people. Dogs such as these had been on their own ever since the first wolves approached the first people.

Interestingly, the Bushmanland dogs didn't hunt, even though hares, rodents, and other small creatures lived near the villages, and even though the dogs appeared to be starving. I asked some of the people why the dogs didn't hunt, but nobody seemed to know the answer, and nobody seemed to have trained them not to. However, the people occasionally hunted the hares and rodents, so, possibly, dogs who did the same were viewed as competitors, and erased from the population. Or perhaps the dogs lacked the energy to hunt. Or perhaps the dogs thought that their chances of finding food were better in the village than in the veldt. If they were away when some scrap became available, they'd miss out and another dog would get it.

I once followed a dog all day to determine his nourishment, and noted that he found and ate two walnut-sized morsels. He also tried to scrape out a cooking pot with his teeth, and may have gotten a few grains of burned cornmeal, but someone saw him and clubbed him with a stick. In the afternoon he went to the borehole where people were watering their cattle. Some of the cows had walked through a wet place, making hollow footprints in the ground that filled with splashed water. The dog drank from these. In the evening he found part of a dry, broken, recently discarded bone from which all meat and marrow had been carefully removed but which he carried into the bushes anyway. I followed him. In the village, he had seemed very shy. He wouldn't meet my eyes, and cringed away when I or anyone else came near him. But he had taken the bone into the bushes so he could have it for himself, and when he saw me coming after him his eyes blazed, his ears shot forward, his hair bristled, and with a roar of a bark he lunged at me, snapping his jaws. The threat was meaningful. I let him be.

In the village, the dogs were competitors. I never saw two dogs forage together. Instead, each searched for food alone, tiptoeing quietly around the edges of the village for a while, then resting to save energy, never finding enough food even for himself, let alone enough to share. Then too, if a dog found a scrap of food, someone would see him eating and, assuming that he had stolen the food, would beat him. The crying of dogs was one of the characteristic sounds of the village. So perhaps this also encouraged the dogs to forage individually. Still, the dogs kept the village reasonably clean, and in this they were useful.

Otherwise the dogs paid little attention to the people except to stay out of their way. The people paid little attention to the dogs except to hit them. People vaguely claimed to like their dogs but affection was never demonstrated on anyone's part—affection for animals was an anathema to these former hunter-gatherers—and it was hard to see why the dogs continued to help the people.

But they did, whether they meant to or not. When strange people appeared, the dogs always saw them first and barked a warning, although most newcomers posed no danger, certainly not to the dogs. The villagers would look up when they heard barking, then greet the visitor, and then would do what people commonly do in all African villages that I've ever seen—they would stone the dogs or savagely beat them with sticks to silence them. Badly hurt, the dogs would cry bitterly and run away, but only to take up another position elsewhere and bark on. To them, a strange person was an invader and the arrival of one was an important event that had to be communicated, whether the people wanted to hear about it or not. And after all, not every human visitor is dependably benign, so these warnings too would be useful.

The dogs also defended the village from wild animals, and in this they showed their greatest usefulness. One night when I was sleeping in the back of a pickup I had parked near the village, I heard a few stressful barks and looked around at the moonlit landscape to see what the cause might be. To my great surprise, a leopard suddenly came streaking toward me, pursued by all the village dogs. The leopard was running as fast as he could, his ears flattened, his long tail aloft, his body folding double each time he reached his back legs forward to launch himself into the next leap. Right behind him came the dogs, grim and silent, ears up, hair lifted, eyes wide, tearing along in a tight pack. Never before had I seen pariah dogs behave with such courage and aggression, but then, never before had I seen pariah dogs chasing a leopard. These dogs meant business and the leopard knew it. He dodged past my truck and streaked off across the savannah with the pack of dogs hot on his heels almost before I realized what I'd seen.

But what *had* I seen? And what had the leopard been doing, to get himself chased so frighteningly? Obviously, he had been viewing the village, probably looking for prey. But for what? A cow? No calves were in the village and adult cattle are too big for leopards, who almost always take smaller prey. A person? Not two months earlier, a grown man in a nearby village had been attacked by a leopard—the man had been sitting at his fire, wide awake, when the leopard sprang on him from behind. The man called for help and his wife beat off the leopard with her digging stick. The man recovered, but had scars.

Perhaps this was the same leopard, trying again. Or perhaps the leopard wanted a dog. Leopards frequently hunt dogs, and the dogs, of course, know of the danger, so they are vigilant. One of them must have noticed the leopard hiding in the bushes and barked an alarm. Surely realizing from the tone of the bark what the watchful dog had spotted, the other dogs then joined together to drive the leopard off.

I will always remember the stirring sight of the six valiant dogs, scarred from beatings, starved and skeletal, grimly chasing a big, healthy leopard in excellent physical condition, who might easily have weighed as much as all six dogs together. Yet the dogs had the courage to challenge him and the will to spend precious energy chasing him at full speed far from the village. And they were doing this all on their own, expecting no help from the people, especially since they chased the leopard too far away for the people to come to their aid if the leopard had decided to defend himself. But then, African village dogs are always on their own, except for one another, and they know this too.

When the leopard and the dogs were out of sight, I listened, but heard nothing. I then waited for the dogs to come back, but they didn't. In the morning I went to the village to look for them, and all six dogs were there. Probably, they had treed the leopard, and then come home. I felt a profound admiration for them. And the benefit of solidarity has never been more clearly shown.

Considering that life in a Bushman village is a fairly typical existence for a Third World dog, and would probably be a paradigm for dogs throughout most of their history, it is perhaps easy to see why dogs don't group together except in emergencies. It is always risky to offer glib genetic explanations, but in this case one might be safe enough in saying that natural selection favored dogs who kept apart, to forage on their own. But what incentive would they have to associate with human beings?

Actually, there was an incentive. By the 1980s, some of the Bushmen were hunting on horseback, and for this style of hunting they used their dogs. Earlier, they had hunted exclusively on foot, stalking the victim. A dog would have been detrimental to a traditional hunt. But hunting on horseback was entirely different, and here, dogs were helpful. The game antelope saw the horse, not the rider, and was not afraid of it, nor was it afraid of the jackal-sized dog, or not at first sight, which meant that the dog could rush the antelope and hold it at bay while the hunter rode up to it and speared it or shot it with a muzzle loader. The dog would get a few scraps from the kill, so it was very greatly to the dog's advantage to help the person.

More important, however, was that the Bushman who hunted with their dogs valued them. They gave them prideful names, and occasionally fed these individuals small amounts of food even if they hadn't been hunting. I knew one man who, on one occasion, purchased a small bag of kibbles for his dog, whose name was Strong. Thus if a dog could find a way to serve a person, or to ingratiate himself with a person, his chances of survival rose. After 100,000 years of this, it is perhaps no wonder that dogs need people more than they need one another. So it's not so much that we seem like pack leaders to dogs. It's that other dogs seem like competitors, and we seem like the source of life.

Rather than forming a permanent group of their own, our dogs formed groups with us, or they tried to. One day, we realized that our main, household group had been divided into four smaller groups, neatly ranked in importance. Each of the four groups contained a dog, and three of them contained a person. All the groups contained a cat or cats. In short, all the mammals in the household had been organized, but not until all the groups were formed, and all their members chosen, did any of the people realize what had taken place.

The dogs, of course, had done this. Perhaps they hadn't formed a permanent all-dog group, as if 100,000 years of life-threatening competition in Paleolithic camps and Third World villages had proved too much for their species, but deep inside their wolfish hearts, they wanted groups, so they made their own. Before we knew it, each of us was seeing far more of one dog than we were seeing of the others, and only that dog would be sure to sleep beside us at night or accompany us wherever we went in the daytime. In short, the dogs had dealt us out among themselves as if we were cards in a deck.

They had also organized the cats. Cats are far more social than most people realize, but in the wild state can't always indulge their social inclinations. Even so, they are grateful for companionship, as pet owners know, and our cats were only too glad to join the dogs. In fact, given the choice, most of the cats seemed to prefer the dogs to the people, and willingly became members of their groups. One day I was working at my desk and Pearl was lying in the denlike knee space under it. The dog-door opened, and a cat came in for a visit. I was glad to see the cat, and expected him to jump up on the desk to greet me. Instead, he went straight to Pearl, who seemed to be expecting him. As he approached, she raised her chin so that he could pass beneath it, which he did, arching his back to brush against her, just as a cat might brush against the leg of a person. He then sat down beside her, and stayed with her for a while. The cat hadn't come to see me. He had come to see Pearl.

The first of our groups was formed by Sundog soon after he arrived. He chose my husband, Steve, and thus Dog One chose Person One, although how Sundog unerringly knew our relative status is less clear. My husband and I would have described ourselves as equals, and didn't have any noticeable division of labor. Our infrequent disagreements were verbal and low-key, and Steve didn't always win them. Even so, Sundog figured us out. Was it that Steve was the larger of us? Did his voice have more authority? Or had Sundog noted our behavior in cars? When we traveled together, Steve usually drove, and dogs know about driving. Had Sundog picked up on that?

Gender also had something to do with the formation of this group. Very soon after this group began, it was joined by Male Cat One, Rajah, and there the membership closed. No other dog, cat, or person was ever included in this exclusive group, of which my husband was merely the focus, not the core, or the controlling member. That role had been taken by Sundog, who knew exactly what he wanted in a group, made all the arrangements, and then kept it that way. He was very cordial to everyone else in the household, as were Steve and Rajah, but cordiality rather than devotion was all that most of us got.

Still, Sundog's group had an importance of its own to the household. Its members were treated with respect by others, and consulted by the others in times of need. Sundog even served as a substitute for Steve when Steve was away, as he often was. During thunderstorms, for instance, the other animals, who ordinarily would have gone to Steve's office to be with Steve for safety, would in Steve's absence go wherever Sundog was and stay with him. Perhaps they weren't members of his group, but when danger threatened, they must have felt that he would protect them. When they felt doubtful, he would know what to do.

It was all very moving. During storms, there Sundog would be, at rest on a rug, and all around him at various distances, cats would be crouching, their front paws folded under their chests, their tails wrapped tightly. They wouldn't be looking at Sundog, nor he at them, but they'd be near him, still slightly nervous, but even so, feeling safer close to Sundog as the thunder rolled. Whatever was making such a noise in the sky might come after a cat, but it wouldn't dare tangle with Sundog.

By day, when Rajah was indoors, he often stayed with Sundog in Steve's office. By night, the three of them slept together, Steve and Rajah on Steve's side of the bed, and Sundog stretched out on the floor beside them. Rajah always chose to sleep next to Steve, if not actually on top of him. Steve, of course, was not exclusive, and wouldn't have denied another animal a chance to sleep on the bed. But Sundog didn't like most of the cats, and would tolerate only Rajah so close by. Then too, Rajah also played a role in the sleeping arrangements. He merely had to be there, his eyes half shut, his ears aslant, his entire being exuding authority, to keep the other cats off.

I was touched to see a sameness in Steve, Sundog, and Rajah. All were male, of course, all were intelligent, and all were the highest-ranking member of their species in the household. But also, all were less than gregarious, preferring one another's company to that of most others. Each of them liked to pursue his own interests without interference or having to share resources, and since the three of them belonged to different species, there was little chance of that. Steve and Sundog didn't hunt mice, and Sundog and Rajah didn't borrow books. None of them paid much attention to the general functioning of the household, but when the need arose, each of them was right there, ready to handle the difficulty. Even Rajah did his part. He would crouch in the open window at night, and spit if he saw a coyote. Even in their sleep, if they heard Rajah's *Ptah!* the other two members of his group would know what he'd spotted. Steve would sit up and shine a flashlight out the window, and Sundog would bark, intentionally rousing the other dogs, who would then make such a commotion that I'd have to let them out, and they'd rush to the place where the coyote had been, all barking. The coyote would be long gone.

17

MISTY also wanted a group, but was less fortunate. When she first came, she wanted to belong with Sundog. When he wouldn't have her, she tried to fit in as a peripheral member with the group he had formed with Steve. For reasons that had nothing to do with Steve and everything to do with Sundog, this didn't work either, so at first, Misty had no one. She could have joined me, of course, and indeed, she spent much time with me, but most of the other animals did too, and Misty wasn't good at competition. She wanted an unchallenged arrangement of her own.

Meanwhile, her heart was with Sundog. Like a wistful high school girl in the presence of the football captain, she had him on her mind. She wanted at least to know he was near, and

was always afraid that he would go somewhere without her. If she heard Steve's car start, she'd look out the window to see Steve and Sundog side by side in the front seat, going away together, with Rajah standing on the lawn, seeing them off. She might run out the dog-door in hopes of joining them, but they seldom stopped to take her along, so she'd watch sadly as they drove away. I often took her with me in my car, of course, but this wasn't quite what she wanted.

After Pearl arrived, she and Misty were amiable enough together and would stay together during walks, but only because dogs like to travel in pairs. Otherwise, Misty spent her days on a rocky outcrop out of the wind where she could see the road. When a car went by, she'd bark once or twice, a sharp and lonely sound, as if she were trying to get its attention.

At about this time, my mother came to live with us, and Misty's fortunes took an upward turn. The apartment, which joined our house through a hallway, had been ready to use for several years, and Misty could have made a place for herself there at any time if she had wanted to, but she didn't. The moment my mother moved in, however, Misty moved right in with her, and there she stayed, no matter who came or went, no matter what else was going on, usually lying curled with her nose in her tail in my mother's velvet armchair.

Misty flourished in this new setting, receiving treats at mealtimes, and pats and praise throughout the day. Because one or another of my mother's companions would sleep in the apartment on a sofa-bed, and because none of them seemed to want to share their sleeping quarters with the animals, they would put Misty out at bedtime. But that was fine with Misty—she would sleep perfectly happily with us, and the moment she was released from the house in the morning she'd run purposefully along her own little, private pathway, not a path that went anywhere straight, but that went behind certain bushes and followed the contours of the house—a path that led to just one place, to my mother's front door, where Misty would scratch to be let in. As her travel had worn the path, so her scratching wore grooves in the door. Every morning was the same. Passionate about her new group, Misty wanted to be in her special chair so badly that she seldom waited for breakfast. Morning after morning, year after year, she wouldn't eat breakfast unless I brought it to her there.

After Misty formed her group with my mother, the only shadow in her life was the arrival of Susan's two dogs, Hobbs and Betty. I was away from home when they first came, but I was told that Misty spent the first week barking. She would follow Hobbs and Betty around, barking bitterly, making everyone's ears ring. But the new dogs didn't use the velvet armchair. Susan saw to that. And after I came home, Susan would leave at night as usual, taking Betty and Hobbs with her, so Misty felt better. Eventually, Misty began to tolerate Hobbs and Betty in the daytime, and they, in turn, didn't try for ascendency over her. They had Susan, Misty had my mother, so two distinct groups coexisted in my mother's apartment, and everyone seemed contented. And just as my mother was glad to let Susan manage the human affairs of the apartment, so Misty was glad to let Betty, the hefty little beagle-cross, manage the dogs. If Don's little Shih Tzu, Wicket, got too playfully rough with a cat, for instance, Betty would woof at him and make him stop.

As the three members of Sundog's group had similarities, so did the group that Misty formed with my mother. For one thing, both Misty and my mother were hard of hearing. Also, both members of Misty's group had secondary status. My mother of course had the highest possible status with all the people of the household, but she didn't make household decisions, she didn't drive a car, and because she functioned among us with soft words, not with labor, the high esteem in which we held her would not have been obvious to a dog. Even so, Misty knew perfectly well who owned the apartment, and it was to my mother, more than to her companions, that Misty devoted her attention.

Misty also perceived my mother's frailty, and saw herself as her protector. When my mother went outdoors for a little exercise and fresh air, Misty would go with her, staying very close by her side. My mother would take one slow step, then pause, then take another step, while Misty copied her exactly, her ears alert, her tail up, her eyes on my mother's feet and on the legs of her walker, taking a step only when my mother did, paying very close attention to my mother's progress and completely ignoring the other dogs who would be running in circles around them.

My mother sometimes recalled the animals of her childhood, one of which had been a pet white rat. She spoke of the rat with such affection that for her ninety-eighth birthday, I gave her a mouse that I'd rescued, unharmed, from the jaws of one of the cats, installing it in a special, deluxe cage with a tower and an exercise wheel. The mouse spent most of his time in the wheel, trying in vain to run off the excess weight he was gaining from the peanuts, cheese, cookie crumbs, and other treats that my mother pressed upon him. The wheel spun, the mouse hurried, and the cats were enthralled. They would sit for hours by the cage on my mother's desk gazing in rapt silence at the active little creature, their ears stiff, their muscles tense, their eyes wide and bright. They were like people watching a thriller video, gripping the arms of their chairs. My mother didn't like this, and would reproach the cats. They shouldn't think such thoughts, she'd tell them, stroking their heads.

Misty took all this in, and one day when a cat deliberately pushed the cage off the desk, Misty leaped from her chair and bounded over to the cat so fast that the cat froze into a crouch, flattened her ears and stared up at Misty, gaping a noiseless hiss. Misty stood right over the cat, looking straight down at him, her tail and ears stiff. Nobody moved. In the undamaged cage, even the mouse kept perfectly still. After that, an entire day passed before the cat tried again to push the cage off the desk.

On another occasion Misty was walking deliberately down my mother's hallway on her way to the velvet chair. The hallway took Misty past the kitchen, where someone had accidentally lit the wrong burner on the stove. A fire had started in a pot. No one was in the kitchen, and no one noticed.

But Misty immediately noticed. She began to bark in an extraordinary voice that she had never used before, very loud and shrill and purposeful, each bark distinct and separate, again and again and again—*Hey! Hey! Hey! Hey! Hey! Hey! Hey!*

There was no mistaking her meaning. Her barking was so forceful, so very different from the ordinary bark which she or any other dog would use if she wanted to go out or saw a stranger, that Susan immediately realized that something was very wrong, came to look, saw the fire and put it out. The emergency was over. Misty continued her careful journey toward the velvet chair.

How could Misty, raised in a crate, understand the significance of a fire? Why had it occurred to her to summon help? Of course she would have smelled the smoke, but why had it disturbed her? She knew all about stoves, and all about things cooking on them. And due to the household woodstoves and fireplaces, she'd often been in the presence of smoke and of open flames. None of these had ever disturbed her. The fire on the stove was different. She immediately understood the danger, and in saving my mother she saved us all.

18

PEARL'S first group member was me. Interestingly, as Rajah had joined Steve and Sundog, so the cat named Lilac, who was Rajah's high-ranking mother, saw herself as belonging with me and Pearl. In fact, we wouldn't have felt like a group without her, as she and Pearl had been close when they lived back in Boulder. Lilac and I were also close, and had been ever since she joined us. As far as Pearl and I were concerned, our group held Place Three in the household—Three rather than Two because everyone is outranked by their mother, and I was no exception. The status situation was different for Lilac, who not only outranked all the other cats, but also outranked the dogs and the people, at least in her own eyes.

As the members of Groups One and Two had certain simi-

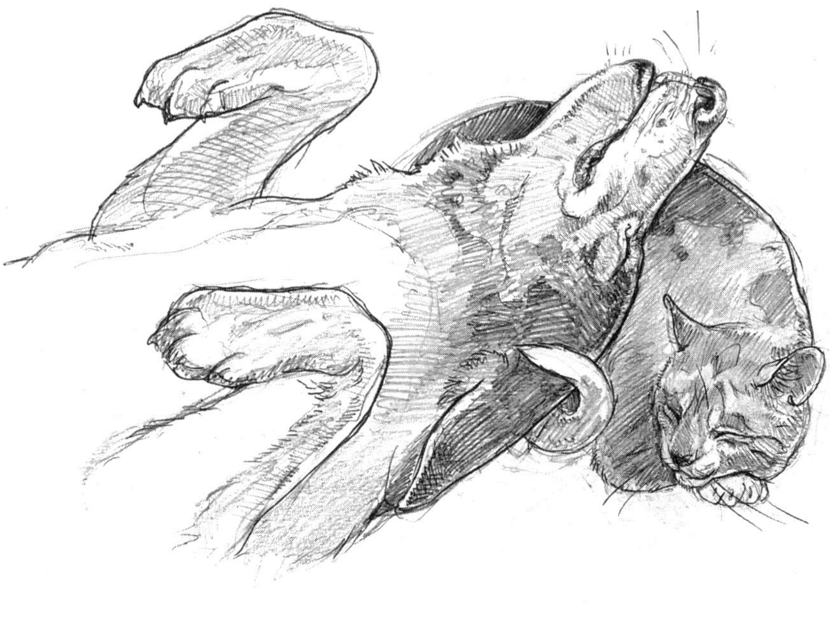

larities, so did those of us in Group Three. For one thing, we may have been somewhat controlling, or so said some of the other members of the household. On my own behalf, since the responsibility for the household fell mostly on my shoulders, I felt I had to be a bit controlling. Pearl obviously felt the same. Hence all the barking, the waking people in the morning, the vigilance, and the protection. She also controlled the cats, whose disagreements bothered her. If two cats tussled, she'd chase them punitively, rushing them to drive them from the room. Yet Lilac surpassed both of us. She didn't hesitate to discipline human beings, as my husband once learned when he shifted his position in a chair and accidentally squeezed Lilac, who, unknown to him, was sharing the space behind him. Lilac gave a ringing call, and when Steve looked around in surprise, she stared him in the eyes, cocked back her right front paw, poised it for just a second so he could look at it, then struck him forcefully with her claws. Whap! *Take that. And next time be more careful.*

Lilac thought that our group should stay together. So did Pearl. Hence my infrequent travels bothered both of them. Neither of them liked to see me packing a suitcase, but where Pearl would merely seem depressed, Lilac would either remove some of the clothes as I put them in the suitcase, or else get into it herself and refuse to get out. She'd tuck her front paws, wrap her tail, and stare at me as I tried to pack. *Don't do this,* her look said. She also would display her resentment after my return. Either she'd climb on some piece of furniture to put herself more at my level, then whack me with her claws as I'd pass by, or else she'd meow to call attention to herself, and then leave the room. I had made her unhappy. She wanted her group at home, intact. My absence had spoiled that.

Our sleeping arrangements were like those of Sundog's group. Pearl slept on the floor on my side of the bed and Lilac curled up between my feet. She was Rajah's mother, and although he could coerce the other cats into staying off the bed, he couldn't coerce her. But sometimes Lilac would lie on Pearl's bed and wrap herself around Pearl's head. Lilac weighed seven pounds and Pearl weighed over sixty pounds, yet seniority is seniority, and Lilac had it. Hence she cradled Pearl as a cat might cradle a kitten, with Pearl's head at her belly, as a kitten's would be. Cats often take this position with kittens, and often treat those who are close to them as if we were their kittens, there being few other paradigms for relationships in a cat's eyes.

Lilac also tried to feed Pearl and me. From time to time, she brought dead mice into the house and put them on the floor, then called aloud. Sometimes she would look at me as she called. If so, I'd think that she meant for me to have the mouse, and I'd thank her, but then after she'd left, I'd throw it away.

One day she brought a mouse and put it directly into Pearl's bowl, which was on the floor. Was this intentional? It probably was. The bowl was always put down in the same place, and had recently been licked clean by Pearl, so it still would have carried the odor of Pearl's saliva. Perhaps Lilac meant the mouse for Pearl, and perhaps she didn't. Again, I threw it away.

Later, Lilac tried again to feed Pearl in a way that I felt was conclusive. I was preparing to feed the dogs and had put their bowls on the kitchen counter when Lilac, again with a mouse in her jaws, jumped up beside me. She then carried the mouse past Sundog's bowl and Misty's bowl and very deliberately dropped it into Pearl's. I didn't want the dogs to eat mice because mice carry tapeworms, so I threw it out. Seeing me do this, Lilac sat down on the counter and gazed at me bitterly for a long time.

But if Lilac was motherly, so was Pearl, especially to the younger cats. She let them play with her ears and her tail, and never objected if they hurt her. In return, they treated her as they treated no other dog. Sometimes when she went for a walk, they would follow her as kittens follow a mother. When she came back, cats would come out to meet her, their tails aloft, as kittens go to meet a person or an older cat. They might arch themselves under her chin, or twist quickly onto their backs in front of her, as a cat sometimes rolls to greet a person. Sometimes when she was sitting on the doorstep of my office, a passing cat would make a detour to greet her. She'd see him coming and would raise her chin expectantly. He'd bow his head and duck under it, arching his back to lightly brush her. He might then lift himself to touch her lips with the little, cleft pad below his nostrils, the pad from which his whiskers spring—the feathery kiss of a cat by which he places a touch of his scent on the recipient. Just what this accomplished, I don't know, unless it was to mark Pearl as his own, in which case, it didn't work—another cat might very well approach her soon thereafter, to greet her in the same manner as his predecessor, and perhaps to sit beside her for a while before going on his way.

Why was Pearl so good to cats? Why was Lilac so good to Pearl? Was this the maternal instinct? If so, I must have had it too, as I spent fully as much time as they did in caregiving. I used to think about this instinct, which certainly seemed to rise unbidden, as if we were vehicles that Gaia was driving. It cost us plenty of time and energy, and our efforts to serve it weren't always wanted, but we couldn't help ourselves.

Perhaps we were like a lioness in Tanzania whom I watched as she prepare to eat a dead, pregnant wildebeest from whose corpse another lioness had already eaten the udder. Lions often start eating at the belly or the rump, and this lioness chose the belly, perhaps because the skin was already open at the site of the missing udder. When the lioness took a bite from that spot, she uncovered a full-term dead fetus in breech position. (Therefore, the wildebeest was probably already in terrible difficulty when the two lionesses found her.) The lioness removed the fetus, but instead of eating it, she began to lick its nose and lips. Gently and carefully, she cleaned its face by removing and swallowing the caul, then cut the umbilical cord, and licked the stump. And there was her mouth, right at belly of the fetus—again a starting place of sorts—so, having done everything necessary to deliver the fetus safely, she ate it. True, it isn't always smart to start eating right at the spot where you've just cut an umbilical cord. If you did, you might eat your infant. But this fetus was dead, and besides, it was a wildebeest, so there was no real reason not to. The lioness might have done the same with a stillborn cub.

Whatever maternal instinct might have stirred in the lioness at the sight of the full-term fetus, its cord uncut and its face veiled in membrane, must also have stirred from time to time in Lilac, me, and Pearl, causing Lilac to nurture Pearl, causing Pearl to nurture other cats, and causing me to stay up late at night making sure that all members of the household were indoors, safe from the coyotes. True, the creatures we protected were adults, or most of them were, but to those of us with strong maternal feelings, the adulthood of children has little significance. Hence as my mother has always been to me, and as I have sometimes been to my adult children (frequently to their displeasure), so were the three of us to the rest of the household group.

Lilac's sense of group seemed to end with me and Pearl. Thus Pearl's sense of herself as a contributing member of the larger, household group was considerably stronger than Lilac's. It was also more complex. How much so was revealed during a little study I undertook to learn more about what kinds of prey the cats were taking. In the process, I found that two of the cats, the cool-minded white cat, Pae, and the marmalade stray named Pula, were making long-distance journeys. Fascinated, I got a little radio collar from a source that supplies wildlife biologists, and made the two cats take turns wearing it. They didn't seem to mind, or to notice that I followed them around the landscape with an antenna, trying to pick up their signals. Pearl usually went with me.

Pae turned out not to be hunting at all, but instead was visiting a neighbor's cat about half a mile away. Evidently her very existence annoyed him. The neighbor's cat was never allowed outdoors, so Pae would harass her through a window. To get there, he'd hurry along the top of a stone wall.

The marmalade Pula was not so easy to fathom. He kept to the edges of the fields, where the woods muffled his signal and I was always losing him. Strangely enough, Pearl seemed to know something about what I was doing. At least, she'd often glance at me when the antenna picked up the signal and began to beep. Then, of course, I would turn the antenna and also start walking in that direction, which perhaps also gave her a hint. But on one occasion the device failed me. Perhaps Pula's batteries had run down, or perhaps mine had, but at any rate, the signal vanished. Now what? The cat could be anywhere. We were standing in an open field on top of a hill, and I was about to give up and go home when Pearl set off in a businesslike manner and trudged down the hill without looking back at me. When she had almost reached the edge of the woods, she stood still, looking intently down at something. Even untrained dogs point—it's how they sometimes show things to one another and also, of course, to us—and Pearl was pointing. She didn't take a stance, or raise her bent foreleg or thrust her head forward as a bird dog might do, but clearly she was focusing, standing squarely on four legs, her ears cupped far forward, and her nose directed straight down at whatever it was she'd found. From time to time she'd glance up at me quickly, then look down again. I went to see what she had found and sure enough, right in front of her was Pula, his faintly striped orange fur camouflaged in the shadowy yellow grass. He was all but invisible. No wonder I hadn't seen him.

Pula had been waiting by a little hole and my arrival annoyed him. He sunk his head down to his shoulders, narrowed his eyes, flattened his ears, and set the tip of his tail to twitching. Maybe the victim had been about to emerge when my footsteps warned it to retreat, and like any other hunter whose moment has been spoiled, Pula wasn't happy

I was, though. I thanked Pearl profusely, and told her she was a wonderful, wonderful dog. She and I went home rejoicing. Pula stayed beside the little hole.

That Pearl's finding Pula wasn't simply a fluke was affirmed later, not once but many times, when she did the same thing again. On these occasions no radio collar was involved—I was simply trying to get all the cats indoors on summer evenings. This had become something of a group activity. I'd go out and call and call, and, hearing me, two of the parrots would also call. They had become so proficient at this that the cats would sometimes come to their voices rather than to mine. When the parrots found themselves getting a response, they took to calling at any time of day, and a cat might obligingly come indoors only to find that she'd been deceived. The more often this happened, the less likely the cat was to come the next time I called.

Thus Pearl's help was particularly welcome. When the time came to call in the cats, she would come to stand solidly nearby. Now and then she'd dash off purposefully, then stand and point. Right in front of her, a cat would be hiding in the tall grass. Intimidated by Pearl's rigid stance, the cat would stay put, and I'd go over and collect it.

I never felt I understood the reasons for Pearl's helpfulness until one winter evening when she didn't help, at which point I felt I learned something very important about her sense of group membership. Ruby, the willful little dickens, had taken off again and was keeping everyone waiting. It came to me that if Pearl would find cats, she might also find Ruby for me, so I got a coat and a flashlight, and roused Pearl from her bed by the woodstove where she'd been sleeping. She got up reluctantly, but, good dog that she always was, she came. To give her the idea of what I wanted, I called and called, then urged her in an encouraging voice to go find Ruby. But Pearl simply looked blank. She didn't get it. And she didn't want to get it. She didn't want to be outside. Cold and bored, she sat down in the snow, looking at me from time to time with lowered ears and rounded eyes.

Actually, Ruby wasn't far away. Soon enough, my flashlight picked up her eyeshine. She'd been lying under an evergreen bush, listening to me calling. When she knew she'd been spotted, she got up and trotted over. At this point, Pearl stood up, sneezed, looked at me, and turned toward the door. *We can go back inside now,* her manner said.

Why hadn't she helped me find Ruby? Almost certainly, she'd known where Ruby was. Dogs see in the dark much better than we do—probably she'd seen Ruby, and surely she would have caught her scent. Pearl wanted to go inside and she wanted me to go inside. That, I believe, is why she sneezed when she stood up. She'd been having strong, conflicting feelings, which at last were resolved. But in that case, why had she simply sat down to wait? Why hadn't she helped me?

At first, it occurred to me that Pearl didn't know I needed help. After all, Ruby was right there in front of me. Although Pearl may have suspected that people can't see well in the dark and have little understanding of odors, surely she had never fathomed the degree of our helplessness. Perhaps in her eyes, no help should have been needed to find Ruby. But then, on the day I lost Pula's signal from his radio collar, Pearl had known where Pula was too. Yet on that occasion, she had volunteered to help me. The two situations seemed to be the same. The only difference was in Pearl's actions.

Then I remembered how, years earlier, Sundog once found

Steve for me. I also remembered that Sundog had declined to look for our lost Fatima, the little dog with diabetes who vanished in the woods. Steve was a member of Sundog's group, but Fatima wasn't, just as Pula was a member of Pearl's group, as were the other cats, but Ruby wasn't.

During the time I was watching the wolves on Baffin Island, I happened to hear the alpha female wolf, the mother, calling from the den site. The four other adults of her pack were all out hunting, scattered far and wide, and she was calling to them. She had a reason to do so—as it happened, she had recently seen something that disturbed her. Within half an hour, a chorus of wolf voices rose from the den site. She and the four other adult wolves were together. Wherever the four others had been, they heard her and came to her call. If an alien wolf, a stranger from another pack, had heard her, he would surely have kept quiet and hurried on his way. Not so with her own family.

So group membership is indeed important under such circumstances, and, I believe, explains why Pearl didn't look for Ruby, and also why Ruby wouldn't come when we called. The wolfish nature of these dogs was telling them that calling, answering, finding, and joining are responses for group members only.

19

FOUR dogs but only three people lived permanently in our house. Sundog had Steve, Misty had my mother, and Pearl had me. No person was available for Ruby. The other group members such as Don and Susan didn't live with us, and therefore weren't candidates in a dog's eyes. And anyway, they had their own dogs, a condition which would have kept Ruby at a distance. For a time, Ruby joined with Don's dog, the fluffy little Shih Tzu, Wicket, in an informal pairing—a play group, really. Together they sped around the lawn and through the fields, rolling on whatever foul thing they found, tussling with each other until they were exhausted, and sharing a dog-bed after that. Wicket would be so tired at night that Don would have to carry him home. When Wicket was gone,

Ruby would go out into the dark and stay there by herself, listening to us call her, but not coming.

As for me, Steve, and my mother, any one of us would gladly have taken Ruby as a group member, and indeed we continually tried to do this, but the dogs who had claimed us kept Ruby at a distance, often with some very subtle maneuvers. If Ruby approached my mother, for instance, Misty would glance at Ruby quickly out of the corner of an eye, and Ruby would lower her head and ears, tuck her tail, and change direction. If Steve invited Ruby to go somewhere in the car with him, Sundog would give Ruby a similar look, and although she'd get in the car anyway because Steve had asked her to, she'd sit as far away from Steve as possible to show Sundog that she wasn't trying to claim him. And if she tried to visit Susan and my mother in her apartment, one of Susan's dogs would give her a stiff, hard look that would dampen her wish to stay. So Ruby would visit briefly, curving her body and wagging amiably to greet all present, but then she'd leave.

Ruby never made much progress in joining the group I shared with Pearl and Lilac. Lilac was sometimes rather stiff with Ruby, staring at her from a distance and walking away if she approached, and although Pearl, the earth mother, was benign toward Ruby and probably never actually tried to repel her, Pearl was nevertheless *there,* always serenely confident that she and I were a unit. Her connection to me seemed total, and this in itself may have been a barrier to Ruby.

I often tried to bring Ruby into the group—I'd take her with me to do errands, I always took her on walks, and I made a special place for her in my office. But my efforts failed. Despite the high esteem in which we human beings hold ourselves regarding dogs, we aren't the ones who make these decisions. All this explained why Ruby was so friendly to visitors, never barking at them, jumping into their cars the moment the doors opened, and then pestering them for attention once they were inside the house. Ruby too wanted a person of her own.

Early on, I realized this, but wouldn't have thought of finding another home for her, or returning her to the shelter, so she could try again. For one thing, we loved her, and for another, not many people want a dog who misbehaves and leaks urine. Let those who do find one of their own, I'd tell myself. The world is full of difficult dogs. Ruby stays here with me.

Besides, Ruby had a group. To be sure, it wasn't the kind of group she was seeking, as it contained no person and no other dog. It was, however, a group in its own right, and what would the others have done without her?

Ruby's group included only cats, but not just any cats. Ruby's group included only herself, the two feral kittens, and Pae, the confident white cat. Why them and no others? At first I assumed that as newcomers Ruby and these cats had low rank and hence, by default, had found companionship with one another. But one day it occurred to me that there might be more to it, and soon became convinced that Ruby and the three cats were together by choice, not by default.

They could have chosen us, the people. True, the feral kittens were terribly afraid of people, but they were terribly afraid of dogs too, and I was making progress with their feelings about people. I'd wait until they were relaxed and sleepy, then I'd pick them up and stroke them until they purred. Soon they began to like this, and eventually couldn't get enough of it. Thus I managed to overcome some of their fears, but they never felt about me the way they felt about the former cat killer, Ruby.

These kittens and also Pae could have chosen Pearl. She was as good to them as she was to the other cats and dogs. Also she would punitively chase any cat who attacked another cat. Hence the shy, low-ranking kittens would have been safe with her, safer than they were with Ruby, who never intervened in a dispute between cats, or indeed, never protected anyone from anything. When danger threatened, such as a very loud noise, Ruby was the first to disappear, abandoning the others to their fate. Yet even though Ruby wasn't protective, Pae and the feral kittens chose her over everyone else in the household. They slept on her bed and often went hunting with her. In every way these humble creatures were as close a unit as was the loftiest in the household, that of Sundog, Steve, and Rajah. One day it came to me that these were the animals who came together from the shelter in Vermont. Perhaps this had to do with their reason for staying together.

The more I thought about it, the more likely this seemed, largely because it fit what seemed to be a pattern. Pearl and Lilac, both from Boulder, stayed together as a unit. Also, three other cats, the brother, sister, and half-sister whom I had adopted from a neighbor, had formed an all-cat unit. But how could this be? At the shelter in Vermont, dogs and cats had been quartered in separate buildings, so that the feral cats and also Pae met Ruby for the first time in my car.

The history of the kittens might have influenced their choice. They were born to a feral mother, who had a nest in the Vermont woods, in a hollow log. There, the mother cat and the kittens were noticed by a passerby, who notified the humane society. Someone—perhaps an animal control officer—set box traps for them, and caught them, or at least caught the kittens.

I don't know what happened to the mother. By the time I came upon the kittens, she was no longer in the picture. Perhaps the trapper hadn't caught her, or if he had, she may have been so wild that she was not a candidate for adoption. If so she was probably euthanized.

In this, our society is exactly the opposite of that of the Kalahari Bushmen, but with equally dire results. The Bushmen believe that animals can take care of themselves, and therefore let them do so as best they can. In contrast, we believe that animals cannot take care of themselves, and are better off dead than on their own. Thus, animals without owners are safer if they live among the indifferent Bushmen than if they live among us, because many animals can, in fact, take care of themselves quite well if allowed to do so, and would like to be given the chance to try.

The feral kittens were kept in the shelter while waiting for adoption, but after they came to live with us, their former life remained very much with them—they spent their days hiding in the basement. At night when the house was quiet they would creep out, find Ruby, step very quietly onto her bed, and curl up in the curve of her body. If a person came upon these three together, Ruby would glance up sleepily, but the kittens would stare wide-eyed with alarm, and if the person continued to look at them, they'd get up and hurry off, keeping close to a wall. They'd come back later, though. Alone in the world, afraid, and without their mother, they were soothed by Ruby's warm presence.

They soothed her too, and also kept her company. Like them, Ruby had also been very much alone. Evidently her former owners had not felt much commitment to her, and had given mixed messages while Ruby was with them. She'd had no other dog to show the way, so she did the best she could, but her efforts got her nowhere with her owners, and she didn't know why. Believing that she should chase cats and kill chickens, she did her best to do so, yet her owners persisted in turning against her. Her place in the household was precarious—that part was clear—but she didn't know how to improve it, and then, one day when at last her owners seemed to take an interest in her and took her with them in the car, she suddenly found herself trapped and alone behind a mesh of heavy wire in a small, bare pen, surrounded by other frightened prisoners, none of whom could offer any comfort.

Death was in the building. From time to time, and under suspicious circumstances, certain animals would vanish. Many animals seem to know about death, so perhaps Ruby guessed what might have been happening. No wonder she rejoiced when she was taken from her cage into a room full of people, one of whom, she hoped, would accept her into a group.

As, in fact, she was. I took her, but only to bring her to a new house with three bigger, older dogs, many strong-minded cats, a few ferocious, screeching parrots, and many alien people who kept watching her suspiciously, in case her urine should leak. So she found a place to live, in a manner of speaking, but it didn't seem welcoming to her.

The politics of our household must have been confounding. Obviously Ruby was going to be Dog Four—the other dogs would see to that—and these dogs would exclude her from the special relationships they seemed to have with the people. Little, low-ranking Ruby wasn't a candidate for inclusion in the many strong and important relationships that obviously existed between people, between people and cats, between cats and other cats, between cats and dogs, dogs and other dogs, parrots and other parrots, and so on through the spectrum of the household, which must have appeared to Ruby as a mass of strangers milling around together in a great, confusing web of social caste, personal histories, overt and repressed emotions, and unfathomable expectations. She must have felt like a shy young woman who comes alone to a glittering party where everybody is rich and sophisticated and knows everybody else.

Such a woman might leave the party, or else she might try to strike up a conversation with someone whose place in the gathering was as fragile as hers, so she wouldn't feel so isolated, so conspicuously alone. Ruby was like that. She tried to improve her situation—she got into other people's cars in the hope that they'd take her with them, and on several occasions she followed a coyote. When these measures failed, she joined herself to the simplest, youngest, least-confident members of the household—the shy little waifs from the hollow log who had been with her in the shelter.

20

THE two feral cats, whose names were Phyllis and Wicca, clung to Ruby. Happy enough when alone with her, they would leave if someone else came into the room. The people frightened them, the other dogs more or less ignored them except for Pearl, who treated them with kindness, the parrots horrified them, and the other cats challenged them. Only one cat seemed to accept them with serenity, and this was Pae, the confident white cat, also from the shelter. He too would sleep on Ruby's bed.

Pae didn't fit the outcast image that was so perfectly embodied by the others of Ruby's group. Poised and assured, he had no problem with any of his social relationships—he managed himself perfectly among the dogs, and was the most pop-

ular cat among our visitors because he was very friendly. Yet like a highly successful person who does not forget his humble origins, he spent his spare time with Ruby and the feral sisters, they who had come with him from the shelter.

One day I happened to notice Pae and Wicca, the black feral kitten, together under the dining room table. To my great surprise, they were mating. How could this be? The people at the shelter had been very specific about the age at which the cats should be neutered, and they hadn't reached that age. Far from it. None of them had showed any sign of spraying or of sexual readiness, for that matter. But obviously we had been mistaken.

"Oh, the hussy!" joked the vet when I later brought Wicca to the clinic to display her bulging abdomen with eight little nipples protruding through her fur. What harm has ever come from an extra cat or two? I planned to keep the kittens. Pae and Phyllis were by then neutered, and we waited for Wicca to deliver.

When the time came Wicca went to her nest which she had made on a sweater in the guest room closet but soon she seemed to be in difficulty. Her tabby sister, Phyllis, knew that all wasn't right, and stayed in the nest beside her, now and then offering comfort by gently grooming her. Ruby also knew, and kept looking up at me, probably to see what I was going to do about it. I took Wicca to the vet and soon thereafter, via cesarean section, she was delivered of two kittens.

The anesthesia had rendered both kittens unconscious. One of them had spinal bifida and was not revived. But the other was in good health. White with black markings like his father, he came home with his mother as soon as she was able to travel, later that same afternoon. At this point, however, something went wrong. Wicca began to bleed, and I rushed her back to the vet, who performed a second abdominal operation to repair her. Thus by evening, little Wicca had experienced a difficult labor, then anesthesia and major abdominal surgery, then more anesthesia and more major abdominal surgery, and she was exhausted and in pain. She didn't get out of the carrying box and didn't accept her kitten, nor did I urge her to do so. Instead, I went to bed in the guest room as nursemaid to the kitten, leaving Wicca beside me in the box where she could see him. The vet had provided me with artificial queen's milk and a tiny bottle, and I tried to do for the kitten what Wicca would have done if she'd been able.

The night passed slowly in the quiet room. Phyllis, Pae, Ruby, and Pearl lay in the hall outside the door, Pearl waiting for me, the others waiting for Wicca, all of them dozing or sniffing at the crack to learn more about the situation, sometimes scratching softly to get in. At about two in the morning I too was dozing when I felt something land on the bed. It was Wicca, who had come for her kitten. As I watched, she lifted it carefully and carried it to her nest on the sweater in the closet. And that was that. From then on, she took care of him.

In the morning, I found a dead mouse outside the door. As most of the other cats were confined elsewhere, only Pae or Phyllis could have brought it. That day, Phyllis insisted on staying with Wicca. Wicca seemed content, more so than she was when she was by herself, so I left them together in the room with food, water, and a catbox, and shut the door. All went well that day. But the next morning, Phyllis slipped out and didn't want to go back, so Wicca was alone. The following day she was restless again, and by afternoon she too was trying to escape whenever the door opened. I tried staying with her to see if my presence would ease her, but it didn't—she wanted out. So I let her do what she thought she needed to do, and left the door ajar, closing other doors so that none of the animals except Phyllis could get to her. Again all went well, with Wicca coming and going as she pleased, but staying most of the time with her kitten. At the end of the day, Pae somehow managed to get through to Wicca's part of the house. He had a dead shrew with him, which he left outside her door.

We tend to think that only mother cats involve themselves with kittens, but this isn't always so. Pae would have known he was the father. Male cats need this kind of knowledge, which is why they have territories—if they can exclude other males, they know that the resident females are theirs alone. If on the other hand a male cat has reason to think that some other male is the father, he may kill the kittens—an act that brings the female into estrus so that he can father her next litter sooner rather than later; thus in no other animal is knowledge of paternity more important.

On the fifth day, Wicca decided to move, as nesting cats often will. She picked up the kitten and carried him downstairs to the living room, to a new nest she had made on a cat pillow that lay beside a desk. Worried that she'd chosen a place that I couldn't defend, I took the kitten upstairs again. But a few hours later he was back on the pillow. Wicca seemed so insistent that against my inclinations I let her have her way. Instead of replacing the kitten in its closet nest, I built a forest of interlocked chairs around the desk so that the dogs, at least, could not get through. All went well for one more day, but the next afternoon I found the kitten dead in the middle of the living room floor.

What had happened? Someone had killed it, probably by shaking it, since I found no blood and no marks on its body. Then I noticed that someone had managed to move one of the chairs slightly. So I suspected one of the dogs—a small one. Alas, this meant that the killer was probably Ruby.

Ah well. Her act was natural enough—some dogs don't realize that kittens are cats or, for that matter, that babies are human beings. Infants, after all, don't give off the same signals as their adult relatives, and in no way resemble them, at least, not to the eyes of many dogs or cats. Hence if one is in charge of an infant of any species, not just a human infant, one cannot be too careful. The death of the kitten was a case in point. I hadn't been careful enough.

Perhaps the killer was Ruby, or perhaps not, but there was no point in trying to determine this, as there would be no more kittens. Pae and Phyllis had both been neutered earlier and the vet had spayed Wicca during the C-section to spare her additional surgery. If Ruby were the killer, she wouldn't have another opportunity to do the same thing again. Feeling more depressed than I had for many years, I carried the little body to the far edge of the field and left it under a tree, where, someday if all goes well, the ashes of my own body will be scattered. In our strangely artificial household, in our mixed-species groups and subgroups, the birth of the kitten had seemed to be one of our very few links to the natural way of things. I wanted the kitten to return to that natural world. I wanted the elements of which it was made to continue in the body of another animal.

When I came back to the house, I saw something that amazed me, something I had never seen before and that once again showed very clearly how little we know about animals. I went into the living room to put the chairs back in their places, and there I found Pae, Wicca, and Phyllis—the kitten's father, mother, and mother's sister—crouching in a circle, facing the spot on the floor where I had found the kitten's body. All three cats were perfectly still and in the same position, their tails wrapped, their legs pressed tight against their bodies, and their front paws folded under their chests. All were the same distance from the spot, and also from one another, and all were facing the empty place. It was almost as if the kitten were still there. The three cats paid no attention to me even after I quietly began to pick up the chairs, and then sat down nearby to watch. Fifteen minutes passed. Then, one by one, the three cats stood up and quietly left by different doors to go on with their lives. I had no idea what I had seen—no idea what the cats had been doing—but whatever it was, I felt sure it had to do with the kitten.

Do cats mourn? We have no difficulty believing that cats

mourn for us, either when we travel, or when we die. They don't want to us to leave them, if for no other reason than that they need us. What we can't seem to accept is that cats might also need and therefore mourn for other cats. But how could an animal have feelings for our species that it didn't experience for its own? Pae, Wicca, and Phyllis were a group that began in a hollow log with the birth of Wicca and Phyllis, to be joined by Pae later, which is exactly how a group of wild cats or barn cats would have formed. Their group was fragile, yes, but when Wicca gave birth, her kitten was a promise for the future of their lineage. All three cats had been ready to support the kitten. Why wouldn't they feel his loss?

I didn't visit the edge of the field where the kitten's body lay because I didn't want the dogs to follow me and find him. Instead, I watched the spot with binoculars, checking on it several times a day. His body was still there on the second day, and also on the third, but in the morning of the fourth day, it was gone. During the night, a coyote or a fisher or a fox had found him.

21

MY HUSBAND is reserved and scholarly, and not particularly social in the usual sense of the word. He speaks several languages but doesn't like to carry on small talk in any of them, so he doesn't much like parties or large gatherings of people where he might be forced to chat. However, he has formed some deep friendships, some from his youth, some from the army, and some from later, and although he may not see these friends as often as he would like, the friendships have lasted for life and even long after—some of his closest friends are no longer living, but he has never forgotten them.

His friendship with Sundog was of this order—deep and forever. Sometimes their bonds seemed eternal, as if in an ear-

lier life they once had traveled together along the edges of the glaciers, as their ancestors certainly did. Once Steve referred to Sundog as "keeper of my soul." From the way that Sundog looked at Steve, ears low and chin high, gently smiling, it was clear that Steve was the keeper of Sundog's soul. Maybe in an earlier life, they were brothers. They were very much alike, as Sundog was also reserved, uncomplaining, devoted, capable, and even scholarly, in his own doggish way. He did what Steve did—going everywhere with him, and never socializing indiscriminately, as most of the rest of us did readily, even some of the cats. I felt sure that Sundog's feelings for Steve were what caused him to share ice cream cones and popcorn in the human manner.

Needless to say, people were much taken with Sundog, especially with his very human ways, so much so that when I described these habits in the aforementioned book on dogs, television crews would arrive from time to time to take pictures of Steve and Sundog sharing. Neither one of them wanted the publicity, assuming they were aware of receiving it, nor did they often watch television—Steve might catch the occasional newscast but normally relied on other sources of information, and Sundog might glance at the TV set if the program in progress included a vocalizing dog, but mostly they regarded television as just one big miasma of inconsequential noise over which the rest of us (the people and the cats, that is) wasted valuable time. And only Sundog really wanted the ice cream, but, because Steve wanted to oblige me, and Sundog wanted to oblige Steve, they both would wait patiently in the hot sun while the camera crew of the moment would adjust the lighting and the equipment, all of which always took a very long time. Sundog found it increasingly painful to watch ice cream drip over Steve's fingers and down into the grass. Finally the crew would be ready and they'd start, perhaps with a fresh cone, so Sundog at last would have a taste or two, but then something would go wrong and the crew would stop the action to fix the problem while Steve and

Sundog waited. They'd then start over. One day Sundog could take it no longer. As the TV crew fiddled with their equipment and the ice cream dripped, he simply grabbed the cone and bolted it. And that was that for Sundog and sharing and publicity. Never in his life did he share ice cream again. Weeks later, after the camera crews had stopped coming, Steve tried again with vanilla, Sundog's favorite, but Sundog grumpily snatched the proffered cone and ate it himself, while glancing now and then at Steve disapprovingly. *You teased me the last time we did this, and I wouldn't have expected that from you,* his glance said.

Some dogs, and also some people, don't do well with publicity. One must really want it to put up with the annoyances, and neither Steve nor Sundog, both ordinarily very patient and obliging, were happy with what they were asked to do for me and my book. Even so, being cooperative group members, both were more than helpful. Without them, the rest of us couldn't have managed. When the dogs and I were asked to appear on one of the morning television talk shows in New York City, Steve and Sundog were pillars of strength in what proved to be a difficult experience.

Our troubles began on the ride from New Hampshire to New York. Very much a country dog, Pearl barked a warning as we passed the first unfamiliar car just as she had done earlier, on the way to Providence. We met very few cars on the back roads, but their numbers increased as we came to the highway, and by the time we were passing through Hartford she was hoarse. Long before we reached New York, she was down on the floor where she wouldn't have to see all the strangers. Her agitation upset Misty, who was carsick.

But if the ride down was bad, the city was much worse—caught in traffic, we appeared to be under siege by armies of strangers—strangers in cars, strangers in buses, and worst of all, strangers on foot, swarming all around us, their frightening faces just outside the car windows, not inches from our own—a country dog's nightmare. Then the hotel, which was very fancy indeed, had revolving glass doors, which none of the dogs had ever seen before, and after these, the elevators. In keeping with the tenor of the hotel, the elevators were luxurious, but the one we entered seemed small to our dogs, who became alarmed when its doors closed behind them. Then, to their horror, it suddenly began to rise. They quickly braced their legs as if to stop themselves, as dogs do when being pulled on a leash against their will. Our dogs could not have been more different from the golden retriever I had seen in Italy, with its smooth, confident behavior in the Milanese hotel.

Sundog tried to seem confident, however. He glanced at Steve, took heart from Steve's secure demeanor, and put on a show of calmness as the elevator rose. Pearl and Misty took heart from him, and when we reached our floor they followed him out the door instead of rushing past him, showing that their panic was under control. True, they pressed against him, glancing suspiciously to the left and right in the ornate hallway, but they trusted his judgment and let him lead the way.

Yet the greatest challenge proved to be bladder relief. Housebroken to a fault, these dogs found no place that seemed suitable. This alarming fact became evident soon after we left the hotel to walk the dogs on leashes. At home, they relieved themselves when they felt like it, leaving the house by the dog-door and finding their own appropriate locations. Leashes, they felt, were for trips to the veterinarian. The last thing one did on a trip to the vet was to relieve one's bladder—far otherwise—and that someone other than oneself would choose the time for this was not part of their experience. On and on we walked, the dogs patiently following, as if they thought we were on an unpleasant journey under conditions that had to be endured.

We might not have grasped the reason if it hadn't been for Pearl. I knew that she wanted to urinate—she hadn't done so for a long time—and had thought that she'd squat as soon as we had negotiated the dreaded revolving door. Yet she did no such thing. Rather, she kept glancing nervously from side to side, cringing away from the many passing people, as if bladder relief was the last thing on her mind. As we passed one of the little trees that line the city sidewalks, each in its few square inches of earth, Pearl noticed an unscooped dog scat. I thought that this might give her the idea, but instead of investigating it, she looked up at me with a woeful expression as if she thought a dog had made a mistake. Then suddenly I thought I saw the city as if through a dog's eyes. To our left was a wall, under our feet was the hard, floorlike sidewalk, beyond it was the floorlike street, beyond that another sidewalk and another wall. At home, wherever she was outdoors, the sky would have been part of every view, the horizon if not the zenith, at least in her peripheral vision. In New York, she saw no such thing. At that, the tiny patch of smog-filled air above the buildings wouldn't have seemed skylike to Pearl, even if she had seen it, which she probably didn't, as dogs rarely look straight up. So she saw neither earth nor sky, and the exhaust-fouled air was the same inside the hotel as it was on the street. In short, she may not have known that we were outdoors. True, here and there a poor little tree gasped for life at the edge of the sidewalk, but it probably did not convey a sense of open spaces any more than would a sickened houseplant, against the overwhelming evidence to the contrary. To Pearl, the city of New York may have seemed like the interior of some giant building.

The more I thought of this, the more sense it made. I had no idea how to communicate the facts of our environment to Pearl, and if not for Sundog, we might have walked indefinitely. But his life probably began in a city—we found him in Cambridge—and no doubt, even after many years, he remembered some of the urban methods of bladder relief. Even so, he too seemed uneasy, although he masked his discomfort better than Pearl masked hers. Unlike an urban dog on such a mission, he made his way along the sidewalk with his head and tail low, and didn't investigate the few markings left by other dogs. It may have seemed to him as though we were traveling to an important destination and didn't have time for him to read the dog signposts as he would certainly have done under other circumstances. He trudged along beside us faithfully, believing that the mission we pursued was ours, not his. In time, however, he must have decided that he'd better take matters into his own hands, and he suddenly veered to the edge of the sidewalk and lifted his leg against a little tree that grew in a grate.

As he did this, Pearl stood still, raised her ears, and watched him with interest. When Sundog finished, she too approached the tree, raised a hind leg, and urinated very copiously, directing the stream against the trunk of the tree. Meanwhile she kept her eyes on mine, as if to learn what I was making of her behavior.

I had never before seen her urinate in that manner, and no wonder. Normally, as everybody knows, female dogs squat, as do male dogs, puppy-style, if elimination is their only purpose. But as Pearl and I looked into each other's eyes, it came to me what she was doing—she was marking. Just as dogs take different positions for marking than they do for ordinary elimination, they also have different rules for doing so. Marking is acceptable where elimination is not, which is why visiting male dogs sometimes lift their legs against the furniture of other people's homes. They are not relieving themselves, they are asserting their status—a very different concept altogether. Normally, leg-lifting in female dogs has to do with estrus, but in the improbable pseudobuilding that was New York, Pearl found no place to eliminate, so she solved the problem of her bursting bladder by pretending to mark. I thought this a brilliant solution.

As for Misty, because she was raised in a crate and taken outside to the dog latrine just twice a day, she could go for eight to twelve hours without emptying her bladder, and she was so upset by the city with its noise and strangeness that no solution to the problem occurred to her. We had to walk her all the way to Central Park, almost a mile away, before she could even begin to think of looking around for a suitable place.

Actually, Central Park was a highlight of our trip to the city. Sundog befriended a group of homeless men who were preparing to sleep on some benches. He walked straight to them, and greeted them with politeness and respect. They also greeted him. One of them even gave him a bite of a sandwich. Often mildly hostile to our guests at home, he behaved as if he had known these men forever. Perhaps he saw them as his hosts. After all, unlike all the other people who were walking quickly past us bound for other destinations, these men seemed at home, in place, and were preparing for bed. As if Sundog felt that he had found the rightful owners of the giant building, he sat down beside them, facing us as if he had joined them, his expression relaxed and his tongue rolling.

Pearl meanwhile noticed a rat under one of the benches, and she chased it. It ran into a bush, alarming other rats, who burst out in all directions. Joyously she chased them all, first one, then another, her fears of the city forgotten. Rats here! Rats there! A rat ran past a group of pigeons, who took to the air with a clatter of wings. In mid-bound Pearl leaped after the pigeons, then dropped back to earth still running. At home, she never chased other animals except to expel them from the property. Here, she was chasing rats for fun. Like a person from the country, dislodged from her normal routines, she was cutting loose in the city and having the time of her life.

If the high point of the trip was Central Park, the low point was probably the television appearance. It went smoothly enough, but the dogs didn't like it—they didn't like the bright lights, or the large number of strangers in the huge, otherwise dark studio, or the confusing activity that was going on all around them. They lowered their tails and ears and stared miserably at their surroundings. No sooner had the program started, with the anchor people cheerfully smiling and reaching out their hands to the dogs, than Sundog stood up and, with quiet dignity, left the set. Too well mannered to show his teeth, too self-possessed to sit rooted in place, trembling and panting, as Misty was doing, he simply got to his feet and smoothly went elsewhere as would any other gentleman who found himself in disagreeable surroundings. His departure seemed like a good idea to Pearl, who was equally uncomfortable and would have followed him, but I caught her by the collar. I shouldn't have. I cherish the video clip I have of Sundog's long, white body stepping quietly off the stage.

Soon after our trip to New York, Steve and Sundog and I went for a walk. Three miles from home, Sundog lay down. He had never done such a thing before, and we didn't know what was wrong, so we waited with him. Obviously, he wasn't comfortable, so when he got up we started home, but soon he lay down again. When we at last reached the house, we made an appointment for him with the vet, and learned that he had serious arthritis in his left foreleg. The condition was so advanced that the vet, and also Steve and I, were astonished that Sundog had not complained of it earlier. He might have walked a little stiffly, but after all, he was fourteen years old, so he seemed normal enough, and when we had started on the walk, he had wanted to join us. But he hadn't limped. Instead, judging from the X-ray, he had endured enormous discomfort without a murmur until he could endure no more.

The arthritis was in the leg that had been shaved when I first found Sundog on my mother's porch—the leg that had undergone surgery. An X-ray showed that steel pins had been set into his bones to hold them together, and the pins were causing the trouble. It was at this time that we learned how sophisticated the operation was, and realized that by questioning the few places where it could have been performed, we might have found his owners. It made me feel terrible to realize this—it was as if for many years we had been keeping something precious and wonderful that didn't belong to us. I even thought of trying again to find the former owners from the hospital records, not to part with Sundog, of course, but to tell the people what had happened to him.

Steve would have none of it. What if the people wanted Sundog back? Although many years had passed since we found him, he might still belong to them. But as far as Steve was concerned, Sundog was ours, and we were his. When it came to our relationship with Sundog, there were laws far higher than the property laws, whatever they might specify, Steve said.

Sundog had surgery to remove the pins, but the arthritis spread to his hips and both shoulders, then to his knees and elbows, then to his wrists and ankles. We tried many forms of medication, which at first he ate readily if we hid the pills in his food, but nothing significantly relieved him. We tried many ways to help him, including putting magnetic pads in his bed, a therapy which some say does wonders for arthritis, but it didn't help, nor did a series of appointments with a certain chiropractor who, out of the tremendous kindness of his heart, treats animals as well as people although legally he isn't supposed to do so. A hush would fall in his waiting room as Sundog entered, walking slowly and obviously painfully but very respectfully and quietly, his head low. All the people in the room were pulling for him.

Of course, we also pursued the traditional veterinary treatments, at the Tufts Veterinary Hospital in Massachusetts, and also with a veterinary surgeon, the region's foremost specialist in bone disorders, but Sundog's condition was inoperable. The surgeon-specialist told me of a recent experience which had greatly upset him. A dog had been brought to him by a wealthy family who drove a new Mercedes that must have cost about $80,000. The dog had hip dysplasia, but would have been a candidate for a new operation invented by the specialist to cure the condition. He was a very nice young dog, the specialist said, and an excellent candidate for the surgery, which promised to cure him completely. But when the owners learned that the procedure would cost $2,000, just 2.5 percent of the cost of their Mercedes, they ordered that the dog be euthanized. The dog was intelligent, loving, friendly, and good-looking, with a wonderful life before him—such a fine example of his breed, in short, that the specialist, knowing where the owners lived, what their professions were, and how easily they could afford the surgery, tried to persuade them to change their minds. They refused. The operation was too expensive, they said. The veterinarian then asked them to give him the dog. He'd perform the operation, and either keep the dog himself or find him a home with someone he knew who would very much like to have him. Again the owners refused. They insisted that the veterinarian euthanize the dog, so he did.

The story upset me so much I felt physically ill. I had a dear friend who cleaned houses for a living. Her husband drove an eighteen-wheeler but was sometimes out of work because of illness. They lived in a mobile home, but had recently spent a small fortune to help one of their two elderly dogs. I felt as my friend did. If money could have helped Sundog, Steve and I would have paid any price to make him well. To think that the wealthy family could have helped their dog so easily! What would we have given to be in their position?

But money couldn't buy Sundog's health. Still, the medication slowed his decline, and he remained active for more than a year. Then the condition progressed to his spine, and partly paralyzed his hind legs. We devised a sling that went under his belly so that we could help him walk outside to relieve himself. But after a while, the paralysis moved to his shoulders and elbows so that he couldn't use his front legs. As far as we were concerned, he could have relieved himself anywhere in the house if he had wanted to, but he didn't want to, so we carried him outside. He learned to relieve himself while lying down. Winter came. He couldn't lie out in the snow, so we made a bed for him in the kitchen where he could be in the center of things, and put absorbent pads under him, pads such as those used in nursing homes for incontinent people. He seemed to understand, and would bark for me to come when he soiled the pads. I would then change his bedding, wash him so his skin would be clean, and put a little ointment on him. I also turned him over frequently so he wouldn't get bedsores. And I massaged him.

When he could still walk with our support, we would help him to climb the stairs at night so that he could sleep in the bedroom. Then Steve had to leave for Europe on business. I helped Sundog by myself. During this time, Sundog lost the ability to walk, and as he weighed more than a hundred pounds, I had to leave him alone in the dark kitchen. But during the second night he was alone, I heard him howl, a heartbreaking sound, and I hurried downstairs to be with him. From then on, until Steve came home, I slept on the kitchen floor beside him.

After Steve came home, we again carried Sundog upstairs together. But his condition had deteriorated further. Being carried frightened him; sometimes he would tremble as he saw us prepare to carry him and the whites of his eyes would show. Even so, he bore his fear like a soldier, as he bore all discomfort. One night we left him alone in the kitchen and waited to hear what he'd do. Hearing nothing, no struggle toward the door, no howl, we looked in at him and saw that he was sleeping. We then thought he preferred being alone to being carried. During the night, one or the other of us would go downstairs to see how he was doing, and offer him a drink of water and a snack if he was awake.

At first, the other dogs sometimes slept beside him. So did Rajah, the cat with whom he and Steve had formed a group. At night, Rajah stopped coming to our room, and in the morning we'd find him with Sundog. Pearl slept with us as usual, but took care to greet Sundog in the morning, when she would kiss his face very tenderly. She didn't want him to be sick. She didn't want him to hurt. But in time, Sundog stopped responding to Pearl and Rajah, and they stopped greeting him. At this point, Steve had to leave again, and I resumed sleeping on the kitchen floor, but now Sundog didn't seem to care whether I was with him or not.

Needless to say, almost everyone we knew was suggesting that we have Sundog "put to sleep" as they so euphemistically called it. To lie in bed, unable to walk, is no life for a dog, they said. But I have always felt that dogs know when their lives no longer have meaning. If they don't want to live, they can always stop eating. And Sundog wanted food.

From the start of Sundog's trouble, I had cooked special meals for him, as he had always loved food, but as time went by, I seemed unable to make his meals a pleasant experience. He'd eat a few bites and then turn away from his dish. It seemed that his medication was upsetting his stomach. The vet added antacids to his batch of pills, and I stopped putting medication in his food, and instead, imbedded the pills in little bits of cheese or hot dogs or butter, but by this time, he had gotten the impression that the food itself was wrong. He'd eat for a little while, then feel sick, and assume that the food was sickening him. He'd stop eating, and would avoid that food thereafter. He also stopped taking his pills. He'd learn, for instance, that a hot dog contained a pill, and would refuse hot dogs. So I separated his medication from food, and shoved the pills down his throat instead. When I did this, he fought me. Otherwise, I offered him new foods almost every day, but each time he thought that whatever he was eating was sickening him, and eventually he stopped eating altogether. I fed him the fortified drinks that are used in nursing homes, also milk and eggs, then just milk. He'd take a few sips, and turn away.

Still, he bore himself bravely, taking notice of all that happened in the room, and asking for food continually, because he was hungry. When I came into the kitchen, he would look at me with great anxiety and bark for food. It was wrenching. He needed food, but he couldn't eat. He needed his medication, but it made him sick, and he may even have thought we were poisoning him. Our veterinarian had done everything he knew to do—he had referred us to every specialist, and tried every drug. Nothing was working, and Sundog wasn't going to get better. We discontinued most of his medication. At this, his stomach felt better and he began to eat again, and soon he was barking for food all the time.

I had felt very strongly that if a dog was eating, he was at least getting something out of life. So at first, it seemed very rewarding that Sundog would eat, but then I began to wonder if his appetite meant something else. People derive comfort from food, and dogs may too. If so, his obsession with food might have meant more than mere appetite. Perhaps eating dispelled some of his anxiety and discomfort. At every meal, when I prepared food for the family, I cooked an extra portion for Sundog.

Meanwhile the other dogs had given up sitting next to him, and wouldn't sleep with us in the kitchen. When bedtime came they could see from my nightgown and sleeping bag what I meant to do, but when I lay down near Sundog they would leave the kitchen and climb the stairs without me, to spend the night, surrounded by cats, in their usual places in the otherwise empty bedroom as Steve had not yet come home. Like Sundog, they too were getting on in years, and were becoming set in their ways, but this surprised me. Not even Pearl stayed with me—Pearl, who ordinarily never left my side. It was as if they had lost all interest in Sundog, or worse yet, were avoiding him.

This gave me a strange feeling. Perhaps they had seen what I couldn't see, or wouldn't see. One day Sundog bit one of my mother's companions, who had only wanted to pat him. He didn't break her skin, and she didn't mind, but his act showed the extent of his discomfort. Soon he was biting everyone who touched him, everyone but me. He didn't warn the person first, perhaps because he didn't realize that he was about to be touched. Instead, at the first pressure on his skin, he'd turn and snap. So it was clear that any touch hurt him. We then saw that his skin was breaking down. The skin on his back was as hard as a board, and the skin on his belly was tissue-thin, both conditions a result of his having had to take so much medication. His discomfort when touched must have been extreme, but, except for biting, he never showed it.

I called Steve in Europe and we discussed the possibility of ending Sundog's life. Steve said that since I was on the scene, I would have to judge the time—he would support any decision. I don't believe in euthanizing dogs, certainly not if they are still eating, but I knew of no way to help Sundog, or to reverse the condition of his skin, or to make him happy, or to restore his health or his comfort. The vet is our neighbor, and had told me that when I felt that the time had come, he'd stop by the house.

Still, I couldn't believe that Sundog's life was over. As he lay on his bed propped up on his elbows, his head high, barking for food which he didn't eat, getting thinner and thinner, biting whoever touched him, I'd wait for a sign. I used to be a Quaker, and the Quakers say, "A way opens." Surely a way would open for me—something would happen that would tell me what to do. But nothing happened. One day for no special reason I made a snap decision and called the vet, and he and his assistant came to the house. Sundog was lying on his bed as usual, propped up on his elbows, but he showed no reaction whatever to the presence of the veterinarian. Our neighbor, Don, happened to walk in at that moment. Don and I knelt beside Sundog and stroked his head—the only place where his skin seemed not to hurt him—and we all behaved in a quiet, matter-of-fact manner while the vet injected a lethal substance into a vein in Sundog's right hind leg. For a moment or two, Sundog stayed propped on his elbows and showed no reaction, even to the prick of the needle. Then he relaxed, and was gone.

This perfect dog was gone, and by my actions. No sooner was he dead than I felt I had made a terrible mistake. Suddenly a huge void had opened in my life, a self-inflicted wound that I knew would never heal. During my life, I had done some very terrible things, and this was one of them. The other people, including Steve, whom I telephoned immediately, felt that the right thing had been done, and the other dogs seemed not to notice the difference. They didn't investigate the place where Sundog had been, or search the house for him. In fact, they went on about their business as if nothing had happened. Many times in the past, I had known dogs to mourn for other dogs. Not this time. Perhaps long before Sundog's death, the other dogs knew he wasn't with them. But he was with me, and I betrayed him. I was very angry with myself and with everyone. When Don came by the next evening, he brought his Shih Tzu, Wicket, and when Wicket approached a dish of food that I had put on the floor for Pearl (another bowl was waiting for Wicket in a corner) I told Wicket that if he touched Pearl's food I'd kill him.

Don looked at me for a moment, then he laughed. "I've watched you all these months with Sundog," he said. "You could no more kill Wicket than you could fly to the moon."

"No?" I asked. "I killed Sundog, didn't I?"

Don seemed shocked. "Don't say that," he said. "You did no such thing."

"Right," I said. "The vet killed him."

"Don't talk like that," said Don furiously. "You did everything you could. You took away his pain. You did him a kindness." So I said no more, but I couldn't change my feelings. I still believe that no dog should be euthanized, no matter how sick he is, unless he no longer wants to eat. Perhaps Sundog wasn't eating, but he wanted food.

Sundog's body was cremated. Steve keeps his ashes. He will also keep mine, if I die before he does, as I will keep his if he dies before me. At that point, we hope that our children will mix our ashes and scatter them at the edge of the woods where the deer come out at night. Sundog's ashes will be part of the mixture. Whichever of us survives the other will already have seen to that. Thus some of our group will stay together, if indeed, in the next world, Sundog still wants to be with me after what I did.

22

WE ASK a great deal of a dog when we take it from its mother as a puppy. We expect it to grow up without parents among an alien species, to learn things its wolf ancestors never had to learn, and to subdue whichever of its ancestral characteristics its owners don't want to see. To know what this must be like, we might imagine the situation reversed, as if, for instance, a human infant were raised by wolves, forced to live in a hole in the ground, eat wolf vomit, drink water from puddles, and kill animals with its mouth. Any child who survived this ordeal would be an international wonder, yet we ask nothing less from every dog.

If the wolves then ran off and left the child to fend for him-

self on the tundra, they would be doing no more than we do when we abandon dogs. Yet people continue to abandon dogs without a second thought, leaving them to live as best they can among the garbage bins of cities. A month or so before Sundog's death, such a dog was abandoned in New Mexico—a little Australian cattle dog–cross whose tail had been cut off. She lived in the streets for some time before she was found by an animal control officer and taken to a shelter. She was only six or seven months old. No one adopted her in New Mexico, but she was such a pretty and promising dog that when her time was up at the shelter, the people at the humane society didn't want her to be euthanized, so they sent her to their colleagues in Boulder, Colorado, to give her a second chance at adoption.

A month or so after Sundog's death, our phone rang. It was our son, calling from Boulder. He happened to be at the shelter of the Boulder County Humane Society, and there he had noticed this dog. The longer he watched her, the more interesting she seemed. In her presence, he took out his cell phone and called me. Here was a dog I should know about, he said. From his description of her dingolike appearance and from his observations of her ability and character, it was clear that she had not been bred solely for her appearance. From the look of her, human interference had little to do with her parentage, and she might even have had a trace of coyote blood. In short, she was the dog of my dreams, the dog I had been waiting for at the time we found Sundog fifteen years earlier on my mother's porch.

By mid-afternoon of the following day, she was on a plane for Boston. I went to the cargo depot of the airline to meet her, and a little after midnight, almost three hours after her long flight had ended, her crate came through the door. Through the bars of the crate, she had befriended some of the baggage handlers and, all smiles, they had been coming to the waiting room to give me reports as her crate progressed through the system. When they opened the crate, she gave them a brief but friendly greeting. Then she stepped out into Boston.

She may have liked the men, but she seemed stiff with me, as if she hadn't known many women, or if she had, she didn't trust them. She was very thin. I had brought water and a bowl, and a box of dog biscuits. I had also brought a leash and a collar, which as she stepped from the crate I put around her neck. She seemed to know what a leash was. Perhaps she had learned that from her sojourn in the shelters. My plan was to take her for a short walk before starting the long drive to New Hampshire.

She seemed very unsure. She wouldn't face me, but instead sat stiffly, glancing sideways as I offered her water and a dog biscuit—she drank some water but wouldn't touch the biscuit—and she wouldn't meet my eyes although I talked to her as we started for the door. She was anxious to leave the building, and tugged hard on the leash.

Once outside, she began a careful, systematic search of the area, sniffing at everything. At first I thought she was looking for a place to relieve herself and I brought her to a grassy strip of ornamental planting. But no—her life as a stray was very much with her. She scraped her teeth on a patch of ice, then licked up the shavings, thus getting a sip of water although the bowlful I had given her was still beside the car. The previous summer, people must have eaten their lunches in this area, because she also found a few ancient food scraps, and ate them greedily. My son had provisioned her for her journey with dog biscuits, but these were still on the floor of her crate. Evidently, she knew more about foraging than about being fed, so it came to me that she might have been a street dog for some time. No wonder she was thin.

On the dark lawn of the cargo building, she never did relieve herself. I was quite surprised by this, as she hadn't relieved herself in her crate either, and she'd been confined in it for at least ten hours. Instead, she searched on and on, pulling at the leash so forcefully as to seem almost frantic. At last I dragged her to the car and we started home, arriving just after two o'clock in the morning. It was a bright moonlit night, and a fresh wind was blowing. All the members of our household were indoors asleep. The new dog stepped cautiously out of the car, caught one whiff of her new environment, and again began to explore very carefully for odors. What she found stood the hair up on her back. Six dogs, all older than she, were somewhere near.

On and on she searched, gathering scent and pheromones, pushing her delicate nostrils into every bush and grass tussock and lingering over every footprint, sometimes with her lips parted and her lower jaw quivering as she used her vomeronasal organ, "doing flehmen" to learn what she could about the many unseen residents of this new place. Whatever else she might have learned, she would have found that the place had been claimed by the six resident dogs—they had marked and overmarked everywhere. But who were they? Where were they at the moment? And what would they do when they met her? With six against one, would they fight?

As if she thought that the more she knew, the better she'd be able to handle whatever fate had in store for her, her search preoccupied her for almost an hour. She examined and reexamined every inch of the lawn, and every now and then she'd bristle, perhaps as she discovered the scent of some new creature. A dog with a group member at the other end of the leash will glance up from time to time to learn if she and the person share knowledge, but this new dog never gave me a thought. With the leash strung tight between us, we went on and on, she bristling and sniffing, and me waiting for her to relieve herself so we could go inside and go to bed. But in all this time she still wouldn't urinate, although at least thirteen hours had passed since she last could have done so. Perhaps she was simply too nervous, or perhaps she didn't dare pass urine and thus appear to be marking, or perhaps she was even afraid to make her presence known. Needless to say, I was no comfort to her. By preventing her from leaving this potentially dangerous place, I was nothing more than a peril, a weight at the end of the leash. She seemed very much alone.

By then, ghosts seemed to move around the shadowy house and over the moonlit fields—the silent presence of all the unseen beings whose scent the new dog was inhaling. Perhaps she was forming images of them. I spoke to her often, of course, trying to reassure her, but she paid no attention to me. Now and then she'd raise her head to test for airborne odors, as if the beings whose scent she found might be nearby. I don't know how long she would have kept this up, but I had been awake since half-past four in the morning as was my custom, thanks to Pearl, so I was overcome with fatigue. I dragged the new dog into the house.

Misty and Pearl were asleep in the kitchen, where they had been waiting for me to come home. (Ruby had gone upstairs to claim the best of the dog-beds, as usual.) Because Misty was hard of hearing, she didn't wake up at first, but Pearl did, and was just starting sleepily to greet me when suddenly she noticed the new dog.

I was ready. I never doubted that my dogs would bark and otherwise challenge the intruder. I also knew that Misty might attack, and although at that moment she was still asleep, I knew that Pearl's barking would have her on her feet in seconds, ready for a fight. She was much bigger than the new dog. I prepared to grab her.

The new dog was also ready. She hadn't wanted this. Surely she recognized Pearl and Misty as the owners of the property—they whose dominating presence had infused the lawn—and here she was, dragged in against her will to confront them. However, she wasn't going to cringe away from the other dogs (a decision she might have made from studying their scent) and she stood on stiff legs ready to defend herself, her eyes wide, her hackles bristling and her ears slightly back, showing herself to be a dog of considerable courage who knew about fighting. I took a firm grip on the leash and got ready to repel her attackers.

Yet what happened next surprised both of us, me most of all. Pearl looked at the new dog for a moment, taking in what was happening. Then she glanced at me—the glance a dog gives a group member to learn if both are sharing the same thought.

Then she approached the new dog slowly and calmly and sniffed briefly at the corners of her mouth. Not a bark, not one raised hair. Pearl didn't even investigate the new dog's anus and vulva, as if she felt that the persona of the new dog didn't matter. In short, her approach to the new dog was the softest possible—the approach of an adult to a youngster whom she already knew well. *I see that you belong here,* her kindly gesture said.

I couldn't believe my eyes. Pearl was the dog who barked at everyone and everything, whether guest or stranger, whether a person, a deer, a heron, a car, a dog, or a cat. Yet somehow, perhaps because the new dog was so young, and perhaps also because Pearl had seen something in my demeanor—something very subtle indeed, as I was not only exhausted but also expecting a battle and cannot have been giving off vibes that were peaceful or good—she immediately understood that the new dog was one of us. Pearl had done something similar on the night she herself first came, also as a youngster in a crate from Denver. Back then, she had walked calmly by the challenging Misty, knowing that she was going to live among us, understanding immediately that she would be Dog Three, and reassuring Misty by her serene and confident manner that all was well.

Pearl then stood politely aside so that the new dog could enter the room. The new dog stepped forward cautiously. At that moment, Misty woke up and seemed happily surprised to see me. Then she too noticed the new dog. The sight brought her to her feet, ears up and hackles bristling. A single bark escaped her lips, but then she glanced at Pearl, who was serene and unworried. Lowering her hackles, Misty also approached the new dog to sniff quietly at her mouth.

If you watch a dog for years on end, you begin to see things. In the few seconds between the time the new dog entered and the time that Misty lowered her hackles, I thought I saw something extremely important. Never before had Misty taken the entrance of a stranger calmly. That in itself was surprising enough. More surprising was that Misty took her cue from

Pearl, not vice versa. So it was Pearl, not Misty, who was providing the leadership, the role that once was Sundog's. His death had left a vacancy at the top of the hierarchy, and it was Pearl, not Misty, who had moved up to fill it. Needless to say, all the dogs had been fully aware of this development, but until that moment, until the arrival of the new dog, nothing had happened to display Pearl's new status to human eyes.

Then it came to me that Pearl may have been providing leadership while Sundog was living. As his health failed and his ability to walk had diminished, she must have taken over, not with force or aggression, but quietly, with kindness and grace. It made sense. Misty couldn't have done so. She was too rigid, too wedded to Place Two. Not only did Misty herself need leadership, but she lacked Pearl's ability to provide it, and both of them knew it. Thus Pearl had occupied Place One in much the same way that a teacher takes charge of a classroom. The teacher doesn't want to suppress and dominate the students—she wants to help them. She takes the leadership because she has more knowledge and experience than they do.

The new dog immediately saw what I had been missing, that Pearl was Dog One, and, after receiving Pearl's greeting rather stiffly, she continued modestly past her into the kitchen. Misty and Pearl turned calmly away as if nothing bad had happened, and the new dog seemed greatly relieved. She had smelled Pearl and Misty on my clothing when I met her at the airport, she had smelled them again in the car on the way to New Hampshire, and she had smelled them powerfully and alarmingly during her intense investigation of the lawn. Their unseen presence had made her very anxious, but when she met them, they accepted her. *So it's okay, then,* she seemed to say. She went past them into the dining room and emptied her bursting bladder on the rug.

I named the new dog Sheilah, not only because she was Australian and female, but because the name Sheilah, with its long vowels, is easy to call. Despite the mishap with the rug, I could see that she was going to be very important to us, another Pearl maybe. She obviously wasn't house-broken, she didn't know her name mainly because she hadn't had one, she didn't come when called, and—as it turned out the next morning—she couldn't go downstairs. Up was easy, down was scary, so I carried her. Nor did she understand the dog-doors. But none of that mattered. She had "dingo" written all over her, and obviously, she'd learn. And she did! She hadn't been with us for twenty-four hours before she knew her name, mastered the stairs, came when called, relieved herself outside, not inside, and used the dog-doors. Not even Sundog would have done better. She was a bright little thing.

There is much to be said for stray dogs. Life in the streets, while harsh and very dangerous, can heighten a dog's awareness to the level attained by wild animals. Sheilah soon proved to be very good at obtaining food, for instance. During her first day with us she jumped up on the dinner table and ate from my plate, a behavior we discourage. Still, for a street dog who knew all about hunger, the act seemed natural enough. What's more, she did this right in front of us with enthusiasm, as if she could imagine nothing wrong. Later, she noticed a parrot drop a piece of cheese on the floor of his cage, and she tried to reach the morsel with her paw. She couldn't. She looked at the cheese for a moment, then caught the edge of the cage paper in her teeth and dragged it out, bringing the cheese with it. As this seemed to be a classic case of problem solving, I gave Sheilah the towel-biscuit test to see how she'd handle it.

Happy to see the biscuit, she cavorted around me in anticipation, but when the biscuit vanished under the towel, she seemed dismayed. So her first reaction resembled Pearl's. She gave me a look that might have asked, *Don't you want me to have it?*

"Go ahead," I said encouragingly. Sheilah looked down at the towel. What to do? Feeling suddenly inadequate in the face of the unknown, she resembled Misty. And, as Misty had done at this point in the procedure, Sheilah gave me another look and left the room. But helplessness was not ingrained in Sheilah as it was in Misty, so she quickly returned to see if there wasn't some way to get the biscuit after all. There it lay, a little lump under the towel. Sheilah sniffed it. Then she nibbled at the lump. So her third reaction resembled Ruby's, who at this point had eaten the biscuit and the towel. But Sheilah was more thoughtful than Ruby, and she knew she didn't want a cloth sandwich, so she stepped back to consider the problem further. Then she sniffed a corner of the towel as if trying to find a way underneath it, nipping the area around the biscuit. Thus she found that the towel moved easily, and she began to shift it about. This almost uncovered the biscuit. Observing progress, Sheilah suddenly got the picture, stepped on the towel to hold it in place, shoved her nose under it and got the biscuit. So in the end, she solved the problem in a classic manner, as smoothly and clearly as Sundog himself would have done.

So Sheilah was smart, she was beautiful, she was capable, and she was good. All this in a dog whose former owners, if any, had done nothing more for her than cut off her tail and turn her out into the streets. She also had a powerful sense of herself, which was partly due to her Australian heritage, and partly due to the fact that, like a Third World village dog, she had managed to survive under adverse conditions, and her ability had given her confidence. She didn't seem like someone who would be content to remain forever on the lowest rung of the social ladder, and I wondered what method she would use to find her way through the byzantine politics of our household to a place nearer the top.

I didn't wait long to find out. Sheilah began with Ruby and also with Wicket, the little Shih Tzu, both of whom, except for her, were the lowest-ranking dogs in our group. They had the habit of playing together, and very soon Sheilah was joining them. She seemed to have boundless energy, far more than either of the others, so she'd play with Ruby until Ruby was exhausted, and then play with Wicket until he had to throw himself down and pant, and while he was resting she'd play with Ruby again.

While Sheilah played with Wicket, Ruby would lie on the grass and watch her in a guarded manner. Ruby was learning that she wasn't as strong as Sheilah. Sheilah was learning that too. Presently Ruby began to give Sheilah more and more space. If Sheilah was standing near a door, for instance, Ruby wouldn't go through it. And at night when we all went up to bed, Ruby wouldn't come. Instead, she hid under the desk in Steve's office, where Sundog used to stay when his legs first began to fail him. After a few days Sheilah took some of Ruby's food, and Ruby humbly moved aside to let her do it. Sheilah was therefore not at the bottom of the ladder, and her arrival had moved Ruby down, not up. Sheilah wanted status, and this was her start.

Sheilah was a very high-powered dog, and although she was young, she wasn't humble. Like the Third World village dogs that she so closely resembled, she immediately saw that I was the person in the household who more or less kept things going, and she wanted an attachment with me. As she saw it, the more firm she could make the attachment, the more likely she would be to gain a solid, safe membership in an important group. She had not forgotten her lonely life as a forsaken youngster in the streets—she knew what it was to be without group support. The first night she spent at our house she jumped up on our bed and lay between us, wiggling with delight at the security this gave her, turning to kiss first Steve, then me, and kicking us with her little round feet. I wondered how Pearl would feel about this—for reasons of her own, Pearl had never slept on the bed with us—but Pearl seemed perfectly content.

However, Sheilah had only come a short way up the social ladder, and I felt that it wouldn't be long before she'd try to move up another rung. It wasn't. One day Pearl was greeting me, and Sheilah crowded in beside her. Pearl resented the intrusion, and displayed her teeth. Sheilah snapped at Pearl, and then began to circle her in an excited manner, meanwhile barking noisily in a sharp, high-pitched voice. She was ready for a fight. *This person is mine now,* her aggressive behavior said.

Pearl was somewhat taken aback. She gave Sheilah a hard look as an adult dog might look at a pup. I spoke to both dogs in a soothing tone, but didn't try to intervene further. Dogs will work out their own social arrangements with or without the approval of the people around them, and the more the people meddle, the more adamant the dogs become. We can no more influence the social arrangements of dogs than they can influence ours, so our interference not only causes additional stress but can also prolong the difficulty. Still, I was relieved to see that for the moment, Pearl decided to ignore Sheilah's display, and Sheilah calmed down.

But Sheilah had not abandoned her efforts to gain status. A few days later, she tried again. Striding purposefully into my mother's apartment, she took a stance squarely in the middle of the room, her head high and her legs braced, and, with "challenge" and "dogfight" written all over her, she began to stare hard at Misty. Sheilah weighed only about thirty pounds, while Misty weighed over fifty pounds and was half again as tall as Sheilah, but Misty was by then eleven years old and Sheilah was just reaching her physical prime. Also Sheilah was fearless and, even more importantly, she had nothing to lose, while Misty was very nervous indeed about keeping her cherished Place Two. Misty became visibly upset by Sheilah's aggressive staring—her gaze faltered, and she seemed about to turn her head aside to appease Sheilah. But too much was at stake. She couldn't afford to appease Sheilah. Slowly, menacingly, she got down out of the velvet armchair and showed Sheilah her teeth. She didn't try a subtle, mild threat with pursed lips, the display that looks like a person saying *shh,* the display that she might have shown a cat, and she didn't use a high-voltage, *take-a-look-at-these-teeth* gape that she might have shown a rival dog when she was younger. Instead, she compromised with a restrained, medium-strength threat, forming her mouth as a person does when starting to say *yes.* The sight of Misty's teeth was all that Sheilah needed. With loud, challenging barks, she advanced.

Her raw hostility intimidated Misty. Keeping her face toward Sheilah, Misty moved sideways to put her head behind a chair, where she cowered and looked out at Sheilah, and then at me. Misty was about to lose Place Two. We both felt it.

Sheilah suddenly spread her legs and went into a squat like a sumo wrestler, the very picture of bold challenge. In this stance, she barked again, now in a loud, triumphant tone, a high-pitched, ringing war cry. She was glad that Misty was cringing. But the people in the room were not. Distressed, they began yelling at her to leave Misty alone. Even my mother stamped her walker and shouted at Sheilah, but Sheilah was totally dog-involved by this time and could have cared less.

As has been said, short of serious fights, it's best to let dogs work things out for themselves, largely because at moments like this their adrenaline is pumping and they don't hear our voices. If the fight becomes serious, one can always break it up by simultaneously grabbing each dog by the back of the neck and jerking them apart, as I'd done many times before, and had scars on my arms to prove it—I had sometimes accidentally dropped a dog, who then lunged at the other dog and bit me by mistake. Bitten or not, I was entirely ready to do it again, but we were in my mother's living room, and the people were becoming upset. So I decided to intervene before fur started flying.

At least, that's the reason I would have given if asked. The real reason I intervened was because I saw that my poor Misty wasn't going to be able to handle the situation. And why should she have to? She spent all her time in my mother's side of the house, and wasn't so much a participant in our side that she would need to hold her place against Sheilah. She and Sheilah didn't really need an interpersonal arrangement. However, if Misty had felt up to it, I would have let her maul

Sheilah a little so that Sheilah would understand who Dog Two was. If right then and there Misty could have halted Sheilah's upward trend, Sheilah would probably have accepted a lower position, if grudgingly, and the social situation would have stabilized for a while.

So for the sake of appearances, I would have screamed and flapped my arms ineffectually while Misty trounced Sheilah, and my behavior would have registered only with the people, which would have been its purpose. The dogs wouldn't have heard me. And soon enough I would have grabbed both dogs and hauled them apart before anyone got hurt. But Misty couldn't carry through. Her lifelong fears were at last being realized. Sheilah was taking Place Two and Misty couldn't prevent it. So I got between the dogs and made Sheilah back off.

Fired with adrenaline, Sheilah didn't care what I wanted of her. As I was reaching for her to carry her from the room, she ducked away, her eyes ablaze, riveted on Misty. Misty was still cringing behind the chair, her face averted, showing her teeth in the *yes* mode and guardedly watching Sheilah, who circled around behind me. With her hard eyes still on Misty, she deftly, quickly, squatted and placed a scat on the rug.

House-broken to a fault, Sheilah was not relieving herself. She was marking. Dogs rarely mark with scats—more normally they mark with urine—yet in such a confrontation, urine means something quite different from a scat. Urine could have said, *I'm just a youngster, and I honor you as my senior.* The scat said *Mine!* Of course, all the people in the room raised the volume of their shouts. I picked up Sheilah and put her outdoors, then returned to clean the rug. Misty seemed humbled. I stroked her. She was grateful. Yet she had lost Place Two to this little street fighter, and we both were saddened by it.

But Sheilah didn't want Place Two. She wanted Place One, and for that she would need to move against Pearl. Pearl by this time was ten years old. She stood seventeen inches at the shoulder and was thirty inches long, not counting her tail. But she was thirty-six inches around the waist, and she weighed almost ninety pounds. To put it plainly, she was fat—the only dog of ours with a weight problem. This detracted somewhat from her beauty, perhaps, but it certainly made her big enough and tough enough to get the best of Sheilah. Yet as far as I knew, she'd never had a fight in her life. She showed aggression, to be sure, but very seldom to other dogs and none to any member of the general household group.

However, Sheilah's behavior would have tested anyone's patience, even Pearl's. One day soon after the episode with Misty, Sheilah came to a doorway near which Pearl and I both happened to be standing. Sheilah tried to run around us, and Pearl snarled. Thus confronted, Sheilah also snarled, and snapped her teeth rapidly several times in front of Pearl's nose. Pearl wasn't going to take that, and with a roar she lunged at Sheilah, who dodged and bit Pearl on the shoulder.

I grabbed Sheilah, so a fight did not develop, and the bite was not serious, but Pearl seemed visibly shaken, and immediately left the house through the dog-door. I found her later in my office, in the denlike knee hole of the desk. She wanted to be alone, or alone with me. She looked past me to see if Sheilah was following, and was relieved to learn that she wasn't. For a long time she looked at me with lowered ears and a sad face. What were we to do about this problem?

That night and for a week thereafter, I learned how badly Pearl felt. She began to have bad dreams. I knew that she was dreaming of fighting, because she would growl and snarl and then cry out in her sleep. In the ten years I had known her and had slept every night at her side, she had never once had a dream bad enough to make her growl, let alone cry. Nor had she cried when Sheilah bit her. On the contrary, she had seemed quite ferocious. But perhaps she had wanted to cry. When she relived the fight in her sleep, I'd get up in the dark and stroke her, and she'd wake for a moment, and look at me softly, giving a few pats of her tail. *Thank you,* her manner said.

It gave me a strange feeling, Pearl's humble, quiet gratitude toward me, there in the dark on the bedroom floor—I had brought in a new young dog to dislodge her from a place she had never sought, but, out of a sense of responsibility had assumed for the good of her group. Now, it seemed, someone didn't want her to have it, and was ready to fight her for it—someone who was in no way qualified to fill it. Pearl was like a dedicated teacher who has been insulted by a student for whom the teacher had always done her utmost and with the best of intentions. The student might want to chase away the teacher, but would lack the knowledge or ability to teach the class. These images came to me as I stroked Pearl in the dark. Short of getting rid of Sheilah, I didn't really know how to help Pearl. The dog politics of the household would boil around me with an energy of their own no matter how much I tried to suppress them. I felt at a loss, unable to solve the problem.

So Pearl solved it. Beloved Pearl, that best of creatures, found a solution that was as brilliant as it was doglike. Its concept was at first unfathomable to human beings, so that we didn't see it coming, and it probably began with dreams.

Dreams are said to be beneficial in that they disentangle our bad feelings, and perhaps this is what happened to Pearl. At about the time her bad dreams ended, I noticed that she was being rather nice to Sheilah. If she met Sheilah in our comings and goings, she would ignore her aggression, and greet her in a friendly manner, not effusively by any means, but in the way she had greeted her on the night of her arrival—calmly, as if she'd always known her, just as she would greet the cats. Soon, I began to notice that if Pearl happened to enter the kitchen when Sheilah was eating, she might help herself to a few kibbles out of Sheilah's bowl. And as she approached, Sheilah would politely take a step back from the bowl to give Pearl access. Female dogs treat their daughters with benevolent dominance as Pearl was treating Sheilah, and their daughters gracefully accept the dominance as Sheilah was accepting Pearl's. In short, Pearl acted as though she were adopting Sheilah, and Sheilah was responding in kind.

Among dogs and wolves and many other kinds of animals

including people, the children of high-ranking parents have high rank too. Thus Pearl was solving the problem of Sheilah. Sheilah was a high-powered dog who wanted significant status, and her adoption by Pearl raised her status. She would have fought for status if she had to, but she didn't have to. Pearl gave it to her, and meanwhile kept her own.

Not every dog would have hit on this solution. Sundog, if confronted by a younger male who challenged his rank, would have shown by his size and his confident behavior that the newcomer's aspirations would never succeed. That might have solved the problem as far as Sundog was concerned, but wouldn't have raised the newcomer's status or stopped him from moving against other dogs. So Sundog's solution, one which many male dogs might have chosen, would not have solved the problem as it applied to the group. Yet if Pearl had chosen to fight off Sheilah, or merely to ignore her desire, Sheilah would not have gotten what she wanted, and our household might have experienced a series of battles, because a low rung of the social ladder was the very last thing on the little street dog's mind. Sheilah might have shown respect to Pearl, but she would have taken on the other dogs one by one to rise above each of them. They, in turn, would not have respected her victories, but would have waited their chance, and then would have tried to regain their losses. Thus Pearl helped us to avoid the blood, the wounds, the endless screaming and crying and the trips to the vet for stitches. There would scarcely have been an end to it. But Pearl took care of that. As Pearl's adopted daughter, Sheilah acquired high status without as much as a snarl or a snap or a wrinkled lip. Then she was happy. It would be hard to think of a better or a more female solution to a problem of hierarchy. But that was Pearl for you. Despite her roughness, her barking, her fearless nature, and her bearlike shuffle, she was a deeply feminine dog. Because of this, because of her, we lived peacefully thereafter.

❖

My beautiful Sheilah! She was as pretty as she was bright, and she was joyous, always smiling. As these dogs often live to great ages, some living well into their twenties, I looked forward to spending the rest of my life with her, as did our other dogs, or so it seemed, after Pearl made her feel that she belonged, after she made a place for Sheilah among us as her daughter. But one day, to our indescribable sorrow, Sheilah died in a mysterious accident that took place very near the house but that none of us saw or heard. No one knew how it happened except the person or people who killed her, and they didn't come forward. Rather, a very kind woman named Anna who helped me keep house noticed her dead body lying near the road where it is crossed by a track that runs through our fields—a driveway of sorts. Sheilah had died just a few minutes earlier.

From the marks on her body, the cause of her death could not be determined, although she had bled from her eyes, nose, and ears, and had bitten through her tongue. However, we felt reasonably sure of two things—a vehicle was involved, but since her body was otherwise intact, she had not been hit or run over. Instead, we believed that something protruding from a vehicle had caught her by the loose end of her collar and choked her. Her collar, which was made of webbing, was found twisted and torn near her body, and her metal license tag was bent.

Everyone except me had been at home. Don was outdoors not a hundred feet from the scene of the accident, building an addition to my office. The rest of the people were indoors, but with the windows open. If Sheilah had cried out, someone would have heard her. A crew of men had been putting a roof on the new construction, and were on their way to lunch in their truck when the accident happened. When Don later asked them if they knew about Sheilah, they said they had no idea. It seemed later that some of the other dogs were aware of the disaster, but if they were, nobody noticed their behavior.

I was doing errands in a town twenty miles distant. I had

finished the first errand—I had bought a lampshade—and was on my way to my second errand when I began to feel a terrible fear about another dog, Ruby. It came to me as clearly as if someone had shouted at me or as if I had seen a picture, that Ruby had been killed by the truck belonging to the roofers. In my mind's eye, I saw the back of their truck with ladders with feet and other hooklike instruments protruding from it. I believed that they had backed up over Ruby.

I had a cell phone with me, and tried to call my husband, but the battery was too low. I told myself that this was nothing, that I should ignore the feeling, but I found myself becoming increasingly frantic from a kind of knowledge so strong and so sure that I couldn't ignore it. Abandoning my other errands, I made a U-turn and sped all the way home, only to see Ruby sitting in a field that overlooks the road. She gave me a frightened look and didn't come to greet me. Very relieved despite Ruby's strangeness, I was about to turn the car around again to return to my errands, when Don came around a corner, and I saw in his face that something dreadful had happened. Then Steve came out of the house and told me the terrible news.

Sheilah's body lay on the lawn, covered with a tarp. Before we dug her grave and laid her in it, I brought the other dogs to see her, so that they would know what had happened. Only Misty seemed surprised, but then, she had been indoors in her favorite chair in my mother's apartment. She sniffed cautiously at Sheilah, her hair lifting. Pearl already knew about Sheilah; she had been outside when the accident took place. Still, she came to look at Sheilah when I asked her, gazed down at her, and turned away. But Ruby wouldn't go near her.

As I brought the dogs one by one to Sheilah's beautiful, broken, stiffening body, and then as I held her in my arms for the last time, I noticed that the roofers were observing me, but when I turned toward them, they turned away, and wouldn't meet my eyes. Perhaps they were responsible, and then again, perhaps they weren't. Perhaps they were, but didn't know it. I looked at their truck for signs of the accident, but if indeed the truck had been involved, the men would have had plenty of opportunity to remove all evidence. Still, what difference did it make? The knowledge wouldn't bring back Sheilah.

Night came. By the light of the full moon Pearl and I went quietly by ourselves to the place where Sheilah's body was discovered. Pearl investigated the ground very slowly and carefully, and what she found made her hair rise all along her back. From her meticulous, silent examination of the area, I could tell where Sheilah had been standing when her collar was caught by something protruding from a vehicle, where she had been dragged, where she had lost control of her bladder, and where she had died. Pearl's eyes met mine, and we gazed at each other for a long time. Then we turned to walk home.

In the moonlight, in the quiet field, I looked up at our house on the hill, remembering my earlier, overpowering sense of something happening to Ruby. I then looked at the place where Ruby had been sitting as I drove up to the house. It was a place she often stayed—she had been there when I left that morning. Presumably, she had stayed there, and if so, she had been there at the time when Sheilah was killed and would have witnessed the accident. In that case, the sense of terror that made me turn the car around and come home was from an image I received from Ruby.

Such was the power of our group. What happened to one of us happened to us all. In our lives, in our minds, in our joys and sorrows, and even in our deaths, we were united. We were related, like the wolves of a pack, or the parrots of a flock, or the cats of a barnyard, or the people of a little band of hunter-gatherers. Perhaps we belonged to different species, and perhaps we had not shared a common ancestor since the Permian or the Triassic, so that three hundred million years divided us, but we were one thing.

Epilogue

No animal is more mundane than a common household pet. There it sits on the kitchen floor, watching you make a meat loaf, just as it has done for the past ten years every other time you made a meat loaf. What is your pet thinking? You know what it's thinking. It would like a little of the meat loaf. Even the naysayers who feel that the minds of animals are impenetrable would have no problem with that.

Yet sooner or later, you will probably have an experience with your pet such as my experience with Ruby and Sheilah, something that will stand your hair on end, something that seems to involve the mind of that animal and your mind too. Without having any idea what happened, or how it happened, you will have exchanged a thought or thoughts with that ani-

mal that are by no means predictable, and in a manner that is by no means obvious. Perhaps you'll somehow know, as I did, that the animal is in some kind of trouble. Perhaps, without tangible evidence, the animal will somehow realize that you're in trouble, or even something as simple as that you're on your way home. You'll be surprised—amazed, even. But if you have any respect for science, you won't mention the event, except to your most intimate companions.

Why not? Because in many respects, the naysaying scientists are partly right—the minds of animals are impenetrable, and ultimately our minds are too. The phenomenon of "just knowing" something about someone else is seldom acknowledged, even informally, and is certainly not understood. That terrible day with Ruby and Sheilah, or that day in my office, when Misty was gnawing at the floor and I imagined myself saying to her, as if she were a person, that she didn't need to do that anymore because she was no longer in a crate, that she was free to go out and play whenever she wanted to, and then she did, immediately, go out to play—these episodes seemed to be part of the phenomenon.

Still, like almost every other communication we have with animals, most of these extrasensory experiences are on such a humble level that they seem not to count. So what if a dog goes out to play? The event is not dramatic, nor can anyone say with certainty what caused it. Yet despite the often mundane level on which we perceive these experiences, clearly something is going on that science may someday acknowledge. A few of my own experiences are presented here because I know that many other people have similar experiences and might like to know they're not alone. They aren't. While I was writing this book Rupert Sheldrake published an entire book on the subject, *Dogs That Know When Their Owners Are Coming Home,** and readers who are interested in the subject might like to read his too.

❖

*New York: Crown Publishers, 1999.

One morning in late November when my husband was away, I was walking to my office with Viva, my young, yellow-nape Amazon parrot, riding on my shoulder as usual. Her feathers had been cut as soon as she fledged, and not all of them had grown back, so she had never flown, and I certainly didn't think that she would do so that morning. But a strong wind was blowing. It lifted her. I felt her claws release my shoulder, and I looked up to see her rising in the air.

I should have cut her flight feathers, as most bird owners do, and as is strongly recommended, for what had just become an obvious reason. Yet I never do this, even though the parrots without flight feathers walk and climb almost as well as they fly, and seem perfectly happy to do so. But a bird doesn't know who she is unless she can fly, so against all advice, my birds were flighted. Even so, flying is like skiing—anyone can point downhill and start sliding, but then what? In the same way, any bird can go up and forward, since this requires only flapping the wings, but flapping won't help her fly down, to bank and turn, to adjust speed, to keep on course in relation to the wind, or to swoop up and land on a perch. All this must be learned before a bird can lie prone on the air like a swimmer in water and shoot forward like an artillery shell, boring a hole in the sky, leaving behind a great turbulence, a circular, windy wake, like a jet plane, which is true flying. However, little Viva's flight feathers had been cut as soon as she fledged, though not by me, and she had never practiced. She didn't feel secure enough to lie prone, but instead, she kept the familiar position she used for walking or perching, with her head up and her tail down. Still, nothing but air was below her, so she beat her wings desperately. Her flight looked like that of an inexperienced chicken who has been tossed upward, and in no way resembled the brilliant, headlong zoom of her wild relatives.

At first, I wasn't worried. I felt sure she'd come fluttering down in the field, and I'd simply go get her. But the strong north wind was under her, and to my growing fright she very quickly flew farther and farther, across a field, across a pond, and into the trees beyond it. There, I lost sight of her. I ran down to where I'd last seen her, a good four-minute trip for a pedestrian, but she was gone.

I then went home to call the police and all our neighbors. I called my husband, who was in Europe, to tell him the bad news. I also called a dear friend, a woman named Sy Montgomery, to whom Viva was very attached. Sy came over, and together we went into the woods and began searching. All afternoon we called and listened, but we found not so much as one green feather and heard only the north wind in the pines.

On the far side of the pond, the woods were very deep. It was already below freezing, and the wind was getting stronger. If this cage bird of the tropics was in a tree, she wouldn't know how to take shelter. Exposed on a branch, she'd get too cold. Or an owl could find her. Or if she were near the ground, a fox or a coyote could find her. Sy insisted that we keep searching, but long before it was dark, when further search would be useless, I felt sure that we'd lost Viva. At last Sy had to go home, and so did I, to round up my cats and feed everybody. I turned on every light in the house, hoping that perhaps Viva would see it, and if she did, that she'd know what it was, so that she could use it as a beacon.

Later I went out again with the dogs and a flashlight. The dogs knew we were looking for something, even if they didn't know what, and they helped me, casting about with their excellent noses, keeping near me rather than running off to amuse themselves as they would if the walk were for pleasure.

I kept calling Viva's name, and then we'd all listen. Nothing. The woods were very dark, the sky was overcast and full of snow, and the wind was biting. At last we went home and to bed, but I couldn't sleep, thinking of little Viva alone, freezing, frightened, doomed. Around midnight I got up and got dressed and with the dogs searched the woods again. Still we found nothing, and at last went home to wait for daylight. Until then, I sat by an open window listening for her voice and thinking of her youth, and her inexperience, and the fact that she knew nothing of flying. Even if she saw the lighted house, or heard me calling, or had all her flight feathers, or was willing to fly at night, would she be able to reach me? That day, Viva would have learned how difficult it is to fly without proper feathers or experience, and the knowledge that she might not be safe would upset her. A few days earlier, she had gone all to pieces because someone had put a pecan in her food bowl. She had never before seen a pecan, and it frightened her badly. How would the dark woods seem to a bird like that?

I was so discouraged that by morning I might have given up, but Sy came back to help me. Sy never gives up, and off to the woods we went again, with field glasses this time, and following Viva's flight line. And almost as soon as we entered the woods, a loud voice called, "Hi, Liz!"

It was Viva, of course. No other animal in the woods knew my name. Somehow, she had miraculously survived the night. We ran toward her voice, calling to keep her answering so that her voice could guide us, until at last Sy spotted her, a little green thing sitting very still and almost perfectly camouflaged, green against green, ninety feet up in a pine tree. We called. Viva answered. We begged and commanded. "Step up!" I shouted at her, which meant to get onto my hand, but, of course, she would have had to fly downward to do it, something that seemed beyond her. So all she did was answer, and only a few times, at that. Then she fell silent and began looking around. Possibly she was trying to think of some other way to get down, or how she would manage if she couldn't. Evidently nothing came to mind, so she looked at us again. Sy courageously started climbing the tree, but it was too tall and the branches were too weak and too far apart. We thought of a ladder, but didn't have one long enough. We thought of the fire department.

Sy stayed in the woods to keep track of Viva in case she flew onward, while I went home to call the fire chief. His name was Steve Black, and he was one of the best-liked people in the community, not only because he was as capable as he was unassuming, but also because of his great kindness, especially to people in trouble. He heard me out, but told me with regret that the fire department no longer helped with pets in trees, nor could they, as the fire engines couldn't get through the woods. Nor was my proposed solution feasible, he said. I had thought to squirt Viva out of the tree with a fire hose, but the only hose that could shoot water so high was known as "the cannon" and its force would reduce Viva to feathers. Discouraged, I went back to the woods to report to Sy, only to see Steve Black coming in from the road to meet us. Perhaps the department couldn't help, but he could—he came to size up the situation and to see if, after all, there wasn't something that he could do personally. This, of course, is why people like him so much.

He looked at Viva and considered the situation, then suggested that we call a tree company and ask for a climber. Brilliant! I ran back to the house to make more phone calls while, again, Sy waited under Viva's tree. Sy had been in the icy woods since early morning, and her feet had become frostbitten. She was surely in considerable pain, but she waited until I got back. Then she went to the house to get warm and wait for the climber while I took her place under the tree.

As the hours passed, the temperature dropped. I kept the field glasses trained on little Viva, and saw that she was looking at me. In mid-afternoon, Sy reappeared, this time with a butterfly net and several men from the tree company, including a climber who strapped spikes on his boots, clipped the net to his belt, and started up the tree trunk.

Little Viva leaned over to watch him as he climbed. She seemed worried by this new development, and the nearer the man came, the further over she leaned to keep him in view. When he was about ten feet below her, he looked up, and for the first time she saw his face. Horrors! He was a total stranger! Viva began to sidle along the branch, away from the trunk, then saw that there was no safe place on the branch where this frightening person couldn't reach her. She had no choice. She flew.

Few sights are as depressing as that of an expensive parrot winging off into the winter woods, especially in the late afternoon. But away among the trees soared little Viva, flying much better than I would have supposed, and heading for Guatemala.

By this time, other people had joined us in the woods, including Sy's husband, Howard Mansfield, two loggers who had been working in the woods nearby and had learned of our difficulties, the forester who had been overseeing the logging operation, the employees of another tree company who had received my messages on their answering machine and had come to offer assistance, and a very kind motorist who had noticed the cars stopped at the roadside and had made his way into the woods to see if he could be of assistance.

The sun went down. The woods turned cold and blue. A blizzard was expected, and just at dusk, the snow began to fall. Howard, Sy's husband, said he'd seen a bird fly through the trees, but it was too dark for him to see what kind of bird it was, or where it went.

Four hours had passed since Viva had been startled out of the tree, and almost thirty hours had passed since she'd left home. When it was too dark to see, I called off the search. I planned to come back to the woods an hour later with a halogen flashlight in case Viva was nearby but too frightened by the activity to make herself known, but in reality, I knew there wasn't much hope, and I tried to get used to the thought that this was good-bye—that I wouldn't see Viva again.

I was extremely grateful to all the kind people who had helped, and I tried to tell them so. That seemed to be the end of it. The people found the road and their cars, and drove away.

I was standing all alone on the road, about to walk home, when something stopped me. I knew I hadn't heard anything except the wind, which was growing stronger, nor had I seen anything in the dark woods, but nevertheless, believing that one should trust one's feelings, I stood there, waiting. And it was good that I did. After a minute or two, a strong but very distant voice in the woods called, "Hi, Liz!"

"Viva!" I shouted, as loud as I could

"Liz!" she called again.

Guiding me with her voice through the tangled woods, she was sitting in a bush not a foot off the ground. I hurried to her, and reached through the branches. In dignified silence, she stepped onto my hand. The pupils of her eyes made pinpoints, showing that she was glad to see me. My eyes must have been pinpointing too. I could have burst into tears.

Amazingly, she was in perfect condition. Her feathers were not fluffed out with cold, and even her bare feet were warm. She also seemed remarkably composed for her experience. Still, she was ready to go home, and, in fact, had turned to travel in that direction. When she left the tree to escape the climber, she flew southwest, so at some point she must have turned around to fly northeast, back toward the house, in order to reach the bush where I found her. The bird seen by Howard had been flying northeast. Surely that bird was Viva.

Might she have found the house by herself? Possibly—the lights were still on from the night before—but she was nowhere near it, and she would have had to batter her way across the pond and the open field, flying straight into the north wind that by then was thick with snow. Even strong, experienced fliers such as wild geese don't often fly under these conditions. So perhaps an angel was watching over Viva. Where do angels belong if not in Class Aves? Who better to care for a bird?

Actually, for all our efforts, Viva saved herself. Or better yet, in the absence of angels, her sense of group saved her. When lost and in distress she called to me, her only group member, and, just as a parrot might have done (although at the time I didn't realize that I was acting like a parrot) I called back. Viva called, I called, she called, and I found her, a parrotlike event. She also called me by name, and in my own language, a call I'd be sure to acknowledge. And I don't believe in angels, much as I'd like to, but perhaps I had actually sensed her presence in a way that I, at least, don't understand.

Viva's was not the only episode with a somewhat unusual component. About two years later, another bird flew away, this time a Moluccan cockatoo, Carmen, whose cage was in my office. One rainy summer morning when I opened the door to my office, Carmen flew past me. She had worked open the latch of her cage door. I called her, but she went so fast that she probably didn't hear me. Then there was nothing more to do than watch her go, knowing from the sinking feeling in my heart that from that moment on, everything would be different.

But unlike Viva, Carmen didn't go very far, and I eventually spotted her in a tree at the edge of a nearby wood lot about sixty feet from the house. Our neighbor, Don, happened to be nearby and he came to help me get her back. But as we walked around the corner of the house, Carmen flew farther away. The wood lot sloped downhill, so Don stood on the high ground to determine her flight direction in case she flew, while I went fossicking around in the woods to find her. At last I caught a glimpse of her and stopped to keep her in sight while Don went near, and when she flew again, he stopped and I went looking. In this manner we looked and called, looked and called, until we reached the road. On the far side of the road was an open field, and while Don searched the woods, I stood at the edge of the field, where I happened to look up at the rainy sky to see a large bird, high and far, flying west. In seconds, the bird vanished in the distance.

It had to be Carmen. She was against the light, and so far off that she seemed gray, not salmon-colored, so I couldn't be perfectly sure, but I recognized her blunt head, her long neck, and her forward wings. And her flight was a parrot's, sure and strong. She wasn't simply moving from one tree to another. She was riding the wind, and soon would be many miles away.

Unlike Viva, Carmen was an accomplished, experienced flyer, whose wing feathers were strong and useful. She could fly up and she could fly down. She could bank around corners, or swoop up to a perch and grab it, so there seemed to be little point in searching any further, or not in the nearby woods. Don went on about his own affairs, and I went back to the house, got out a map, and called the police in every town in Cheshire and Hillsboro counties. I then wrote out a description of Carmen that included the ID number of her ankle band, and faxed it to these police stations. I also got out the Yellow Pages, and phoned and faxed every pet store and veterinarian listed. There is much to be said for the human race, or at least, for those members who concern themselves with animals—every person I spoke with was willing to help. This took many hours, however. When I finished I went out in the rain to search and call again. But I felt that this was useless. The sky seemed huge, and except for the rain, the woods were empty. Carmen was very far away.

For the rest of the afternoon I waited for the phone to ring. Every time it rang, I told the caller that I had to keep the line open, that I'd call back. Soon I began to feel that I wasn't going to get any news of Carmen, and I began to tell myself stories. She'd be fine. It was summer. The snow was many months away. She'd find sunflowers and eat the seeds. Lots of people grew sunflowers. Or she'd find a bird feeder. Everybody had a bird feeder. The owners would see her, and would take her in. And they'd like her. All cockatoos are wonderful, but Carmen was a dream. And she was beautiful. Obviously, she wasn't local, but was a special, exotic bird, and whoever found her would know that. She was also big, as big as a big hawk, too big for a hawk to tackle. And she knew about hawks. All birds seem to know about hawks. She even disliked the silhouette of a hawk I had pasted to a window. The first time she saw the silhouette she threatened it. Later, she pulled little pieces off it, perhaps so that it wouldn't look so disturbingly hawklike. She was smart, and she was ingenious. She'd be okay.

Drained of all emotion, and all but resigned to the loss of my Carmen, I opened a book and began to read. Soon I was absorbed by it. The sun must have set. The room grew dark. I got up to turn on a light, and realized how quiet it was. The dogs were sleeping. The windows were wide open, and I could smell the rain. I found my place in the book and continued reading.

But suddenly, I began to feel something, perhaps like the feeling I'd had just before I'd found Viva but much stronger, something very hard to explain. The feeling came slowly, but kept growing, like the light that slowly builds before the rising moon. Carmen. I knew I was feeling Carmen.

At first, I told myself to stop this. Was I supposed to jump up every time I thought of her for the rest of my life? I went on reading. But soon the feeling was so strong I couldn't see the page. So I went outdoors into the rain and the heavy dusk and stood on the grass, waiting quietly. Carmen was near. I knew this without any question. And in less than a minute, she called.

I answered. She called again. I ran in her direction, still calling because at first, in the thick woods, in the dark, I couldn't find her. She answered immediately every time I called, and soon I saw her shining on the low branch of a tree. I walked up to her, holding out my hand, and, first one, then the other, her warm feet gripped my fingers.

I have no idea what to make of this. I have no idea what happened, or where she went, or how she found her way back, or how I felt her presence. What can you do in a case like that? I got a better latch for her cage.

❖

The experiences with Viva and Carmen were not the first of their kind. Something similar once happened between me and my mother many years earlier, while she still lived in Cambridge. I was involved in a small research project in the basement of the nearby Peabody Museum at Harvard. One night I worked there very late, long after the building had closed. The basement seemed to be full of mummies, which I didn't especially like, and I didn't think much of the jars of body parts pickled in formaldehyde on the shelves, especially the severed head of a proboscis monkey from which the pickling solution had evaporated, but I was okay and held my uneasiness in check until, suddenly, all the lights went out. Panicked, I groped my way in pitch darkness through the basement, past the jars and the mummies, up the stairs, and through the empty, echoing halls, toward the front door. As I fumbled around in the blackness I heard slow, measured footsteps coming down the hallway, as someone or something advanced on me. I'm ashamed to say that I was so scared I thought my hair would turn white from sheer terror.

Of course, I was not in real danger. I learned later that the footsteps were those of the janitor, a kindly and reasonable person. Probably, he heard me rustling, and since he hadn't known I was in the basement when he turned out all the lights, he was coming to see what was making the noise. But I didn't wait to find out. The moment I got the door open, I bolted at full speed to my parents' house.

To my surprise, my mother was standing on the front steps looking for me, although I hadn't told her I'd be coming. She was very anxious. A few minutes earlier, she had gotten a powerful feeling that something terrible had happened to me, and she had come outside.

Well, something terrible had happened. I'd been scared half to death, not a quarter of a mile from her house. Normally in stories of this kind, the person is in actual danger. I wasn't. But I certainly thought I was, and if indeed she received a thought from me, that was the one she received.

❖

A few years later, Steve and I took our children with us to live in Nigeria. We also took our dogs. One of these was a little pug with negligible house-breaking skills. One afternoon, when we were about to go to a wedding, I came into the bedroom to get dressed, only to find that the dog had puddled on the floor. Steve was lying on the bed, dozing, and the dog was looking at me. That was it. Nothing else happened in any observable way.

However, I began to have a fantasy. We'd go to the wedding, I imagined, and we'd come back to find messes on the floor. Not wanting this, I imagined putting the dog outside on the balcony off the bedroom where she could puddle as much as she liked. To this end, I imagined myself crossing the room, opening the bureau drawer, taking out a set of keys, unlocking the balcony door, and going out to the far end of the balcony, followed by the dog. Picturing myself with the dog at my feet, I imagined looking around the neighborhood, and saw in my mind's eye what the neighborhood looked like from there. In my fantasy I planned to leave the dog on the balcony, go back to the bedroom, and shut the door. It was hot. I began to think about a bowl of water and perhaps some shade for the dog. But while, in my imagination, I was still outside with the dog beside me, my fantasy ended. I was irritated with myself for even entertaining such an idea. I couldn't do a thing like that to this little dog. She would get overheated and be miserable. I'd have to put up with the messes.

All this time I hadn't moved or spoken, but just stood still in the middle of the bedroom, thinking these thoughts The dog hadn't moved either. On the bed, Steve said, "Don't leave her out there."

Oh, wow. "I won't," I told him, "but it's strange that you should mention it."

How had Steve known what I was thinking? He said that while half asleep he'd heard the sounds of my progress, the opening drawer, the jingle of the keys, the handle of the balcony door, the dog's footsteps clicking on the tile floor. He thought I'd taken her outside onto the balcony.

Well, that's what happened.

Nothing could be more abhorrent to those who follow science than theories about psychic messages, and rightly so—the entire subject seems improbable. Nevertheless, improbable or not, the so-called psychic events continue to occur, and supernatural explanations are certainly tempting. But just because the events seem supernatural doesn't mean that they are. Some are later found to have very earthly causes. For instance, psychic powers were tentatively and perhaps jokingly offered to explain how elephants seem to know what other elephants are doing far away. The real cause was discovered by Katherine Payne in 1983 when she demonstrated that elephants make far-traveling calls too low for people to hear.* The elephants were communicating with infrasound.

Perhaps emotional changes might engender certain odors, however faint. So perhaps some of the seemingly psychic insights might be attributed to the vomeronasal organ, fully functional in most mammals, now thought to be functional in people too. In us, its opening is in the little lump in the roof of the mouth, and it detects some of the chemical signals, the pheromones, that rise from such sources as other people's bodies. Unlike sights, sounds, tastes, and odors, however, the information gathered by the vomeronasal organ doesn't seem to pass through our conscious minds. So perhaps there is indeed a sixth sense, and if so, we all have it, not just the psychics. Still, an ambient chemical in the environment, whatever its source or its meaning, seems an unlikely explanation for what happened between me and Carmen, or between me and my

*K. Payne, W. Langbauer, and E. M. Thomas, "Infrasonic Calls of the Asian Elephant, *Elephas maximus,*" in *Behavioral Ecology and Sociobiology* (1986), 18:297–301.

mother, or to me, my husband, and our pug dog that afternoon in Nigcria.

So who knows? Perhaps the latter episode was a coincidence, the preferred and often accurate explanation for events of this nature. But that wouldn't have been likely—perhaps my sleeping husband had known of the puddle through olfaction, but he had absolutely no reason to think I'd leave the dog on the balcony. I had never done such a thing, and never would leave a dog imprisoned in the African sun, for any reason, as he well knew.

Or perhaps a thought really can fly from creature to creature. After all, the faint, electric disturbance caused by thoughts can be detected outside the skull, so they might find their way to someone else's head, assuming they were strong enough not to get lost among all the other such disturbances in the environment. How is it, for example, that we sometimes sense when another person is looking at us? No thought to that effect passes through our minds—we simply look up, and meet the eyes of the other person. Nor do we need to know the person. In my case, the phenomenon occurs most frequently in traffic, when I sense an observer in another car. So powerful is this effect that certain hunters, whether human or animal, are careful not to stare at their potential prey. When stalking, one does best if one can watch the intended victim with one's peripheral vision. Lions certainly seem to know this. A lioness in Namibia who was stalking me did. Knowing that she might be in the vicinity, I had been looking around for her with a flashlight when fortunately I noticed her advancing. She kept me in view out of the corner of her eye, being careful not to look at me directly. Obviously, she botched the job as I'm here to tell the story. (Good thing I had a flashlight.) Meanwhile, the fact is, we know so relatively little about these peculiar phenomena that, at present, we can merely guess as to the mechanisms that may prompt them.

❖

If I were a scientist, I'd be obliged simply to drop the subject at this point and avoid it forevermore. But I'm not a scientist. I am merely a chronicler with a respect for science. And in that role, I would describe a fourth event that happened one gorgeous summer day when Steve, Sundog, and I were in the parking lot of our local shopping plaza. I personally felt sure that some kind of unexplained messages seemed to be passing, and, remembering the experience in Nigeria, I began to wonder if the vector was sometimes a dog.

Steve and I went to different stores to do our errands. I didn't know what Steve's errands were or how long they'd take, so when I finished mine I went back to the car to wait with Sundog, who, because it was hot, had gotten out of the car and was sitting in its shade. I sat on the hood, glad of a chance to relax, feeling very happy and enjoying the glorious day. But suddenly a sense of utter sadness engulfed me like a tidal wave. It was as if the sky had suddenly turned black and the day had turned cold. Everything good turned to ashes. My life no longer had meaning. I began to cry. Embarrassed to be weeping in public in the parking lot, but having absolutely no idea what had come over me so quickly, I looked down at Sundog for consolation.

But he had his back to me and was looking very intently at one of the stores, a travel agency about fifty yards distant, on the far side of the parking lot. Sundog's head was high but his ears were lowered, so that he looked like a dog who is seeing something dreadful. I followed his gaze but saw nothing in particular. He continued to look at the travel agency, however, and presently I saw Steve come out the door. Steve too seemed downcast. What was wrong?

Steve told me that the travel agent, Bernie, had asked about our daughter. Steve had no idea that Bernie knew our daughter. But he did. More than fifteen years earlier, our daughter was in a tractor accident that left her permanently paralyzed. Bernie explained that he had driven the ambulance that came to collect her. When he met our daughter, so to speak, she was lying half-dead on the road. He was the first person to see that she was paralyzed, and by taking the appropriate actions, he saved her life. Steve had never known this about Bernie, as he was away from home at the time. Nor, in the midst of the crisis, had any of the rest of us learned the identity of the ambulance personnel, nor would we have known them at the time.

In the coming years, our daughter went to college, got a good job, married a Vietnam veteran, and moved to Texas, where she and her husband became activists on behalf of civil rights for the disabled. Although as a result of her work she'd been in more jails than Al Capone, President George Bush invited her to the White House for the signing of the Americans with Disabilities Act. To think about her made us happy, never sad, and anyway, a wheelchair is by no means the disadvantage that some people seem to believe. For many, many years, Steve and I had scarcely given the accident a single thought.

But that day in the store, when Bernie told Steve how he had met our daughter, Steve experienced anew the emotions he had felt upon learning of the accident. And so did I, sitting on the car in the sunshine.

Okay, I can't explain it. But I'd say this much: I doubt that the laws of chance explain it either. During the many years since the accident, both Steve and I visited the travel agency often, and we both had come to know Bernie well. But only on that one special day in the parking lot did the subject of the accident ever arise with Bernie, and when it did, my husband and I both relived that terrible day. Or Steve relived it. I relived only the feelings that resulted from it. And if you ask me, Sundog was the vector. If indeed he had caught Steve's emotion, he did so while Steve was out of sight, inside the store. But all he caught was the emotion, since he couldn't have understood the accident itself, which had happened before he was born. Perhaps that's why I experienced only the emotion, not the memory. If indeed I caught the feeling from Sundog, the feeling was all he had to transmit.

I sometimes read about ESP, often in the *Skeptical Inquirer,* whose contributors debunk it. And reasonably so. Many of the claims of the so-called psychics are both wild and irresponsible, and would not be credible even if normal rather than paranormal explanations were offered. I also read of an experiment to test for ESP in which a subject is shown a series of cards with geometric images on them, and is asked to transmit the images via ESP to someone else in an isolation booth. The experiment suggests that transmission does not take place, to no one's surprise, certainly not to mine. But is the experiment reasonable? I personally feel next to nothing at the sight of a geometric image, but I feel a terrible and powerful emotion about a serious accident to my daughter. Maybe if the material to be transmitted were less boring, the experiment would have different results. Maybe emotion, that bugaboo of scientists, must be present for transmission to occur.

Perhaps ESP exists, or perhaps it doesn't. If it does, it would be hard to imagine a better way of transmitting information without mail or telephones or, in the case of animals, without speech. In the Kalahari in the 1950s, the Bushmen relied on it although they were certainly not a superstitious people. They had a practical explanation for almost everything, and they were usually right—at least, many of their views agreed with those of scientists who came later to study them. Laurens Van der Post wrote of Bushman ESP in his book *The Lost World of the Kalahari,** an account I have always taken with a grain of salt, since the surrounding circumstances are downright improbable. He claims that a group of Bushmen, knowing in advance that faraway hunters had killed an antelope and were bringing the meat to their village, began to sing a song specific to the animal they had killed. The trouble with this story is

*London: Penguin, 1968.

that the choral singing of Bushmen is not used as an expression of joy, certainly not with the music pertaining to that particular kind of antelope, as this music is reserved for matters involving menstruation. A comparably unlikely story about our culture would be that some people mystically realized that a pizza would soon be delivered, and spontaneously celebrated a high school graduation or baptized a baby as if that were the appropriate response. Some things just don't happen and some accounts just don't fit.

Even so, the Bushmen whom we visited took certain dreams to be messages, especially if the dreams seemed to bring news of family members who were far away. They saw these dreams as we see phone calls. Sometimes the news brought by a dream was accurate, and sometimes it wasn't, but the Bushmen always took it seriously. Interestingly, though, while dreams about other subjects were also taken seriously, and were interpreted in different ways for different reasons, they weren't accepted as actual news. To dream about a rainstorm, for instance, would not mean that rain was coming. Instead, it might mean that the power of the rain was lurking, and that if people were to dance, they could access that power and use it to benefit their group—not to bring rain, but to bring something less tangible—the resolution of a quarrel, for example. But to dream that one's nephew had died would mean just that—the nephew was dead. The dreamer and others would weep and mourn. Among people who have no way to communicate with distant relations other than by making long, dangerous journeys on foot, but who nevertheless care deeply about one another and think continually of one another, a dream of this kind can seem very important.

When we lived in Nigeria, Steve and I often encountered people who had experiences that would not be credited in the United States. For instance, two famous holy men from Nigeria's northern region correctly diagnosed a fatal blood disease in a British journalist just by glancing at him. They had never seen him before—they had never even heard of him—and he himself had no idea that he was sick. Nor had he come to them for medical reasons. Rather, he had gone to interview them about a political situation and they had taken one look at him and divined his fate. African science, the diagnostic method was called.

Whatever is going on, we in the West are stumped by it. We obviously have no idea how to apply the scientific method to the subject. Perhaps this is because we misperceive the phenomenon, clumping it with ghosts, crystal power, contact with aliens from outer space, astrology, channeling, reincarnation and the like. If so, we'll never understand it. Or perhaps we are approaching the puzzle from the wrong direction. Perhaps by insisting that all emotional content be removed from the experiments, we are removing the very element that could bring a positive result. But someday, someone may change all this. Someone may notice something common to ESP experiences that will open a door to the phenomenon. Perhaps this person will be a scientist or perhaps not, but whoever it is will be a good, perceptive observer, someone who notices everything, who finds it easy and natural to pay very close attention no matter how humble the subject or how small the details—in short, a person who could rival the observant horse of Germany, the horse for whom no signal was too subtle—the famous Clever Hans.

Appendix I

On the Control of Dogs

BASED on experience from having written an earlier book on the natural behavior of dogs, as a result of which I received more than 160 gallons of mail, tightly packed, I learned that many readers would have preferred to learn about dog training. I am not a trainer and cannot offer technical advice, but I do have a few hints. I submit them here not only for the sake of dog owners, but also for the sake of dogs, because every hour of every day, perfectly capable people are abandoning perfectly good dogs merely because the dogs have behavioral problems which their owners misunderstand. Most of these problems are correctable, yet many of the methods which certain people recommend are not only useless, but damaging. This springs from an unwholesome attitude quite prevalent in this country about the need to rigidly control dogs.

A castrated or spayed purebred dog who was taken from its mother at an early age, raised entirely by our alien species, and trained to excel in an obedience trial is the paradigm for correct dog management. This life is as unnatural as that of a circus elephant. Yet dogs are expected to adjust to it, and most of them do, often so subtly and inconspicuously that we are unaware of what they're doing. The credit is not ours but theirs.

Some dogs, however, do not make the adjustment. Nicholas Dodman and others have written very informatively on the behavioral problems of dogs, pointing out that unwelcome behavior is by far the greatest cause of death in dogs—not disease, not overpopulation, not runaway strays, not indiscriminate breeding in back alleys. People acquire a dog, don't understand it, can't train it, get fed up, and take it to the local humane society where they offer it for adoption, hoping to pass on the problem to somebody else. But nobody wants a problem dog, and the shelter is already overrun with similar dogs. For the vast majority, new homes cannot be found, and the unwanted dogs are executed. Or "put to sleep," if that sounds better. If a virus were killing dogs at this rate, says Dr. Dodman, the nation would be up in arms to find a vaccine.

My first bit of advice, therefore, would be to read *Dogs Behaving Badly* and also *The Dog Who Loved Too Much* by Dr. Dodman.*

My second suggestion would be to relax. Dogs are extremely sensitive to human tension, especially when they themselves are the cause. Nowhere is this more manifest than in housebreaking. Imagine yourself as a rather small animal in the presence of a very large one. The large animal has tremendous power over you—he is your only source of food and shelter (you can't eat until he wants you to eat and you can't get in or out of the house unless he opens the door). But you can't seem

*Both published by Bantam.

to please him. He seems always to be fuming, and the trouble has something to do with you. You don't know what or why. You suspect that the problem may be your urinary habits, but you don't know what you're doing wrong. Are you marking the wrong places? Should you keep to the living room rather than the hall? Did you fail to overmark an earlier stain? Are your marks too high, too widely spread, as if you overestimated your rank? Or are they too narrow and small, and thus inadequate? Should you refresh your marks more often? He himself marks inside a big white bowl (no wonder he's worried about marking—his own has no effect), so when you yourself visit that bowl to drink, you find no residue to suggest his preferences. Even so, he's always angry, always looking at you with suspicion, all of which makes you unhappy and jittery, so you drink and drink to calm your nerves, and then your bladder needs relief. Trembling with fear of the large animal's capricious and volatile anger, you sneak up onto a bed which will absorb the telltale puddle, and you relieve yourself there. The next thing you know you're in a wire cage in a strange building, where you wait for a few days, alone and frightened, until a stranger in a white coat hauls you out and sticks a needle in your vein. The room spins, then everything goes black, and you are history.

We human beings have great difficulty learning things that to us seem meaningless, such as long strings of numbers, for example. Many of our requests seem meaningless to dogs, at least at first. Dogs have their own dog rules about where to eliminate, which makes the process important to them, not only for personal relief, but also for canine social reasons. It is our task, as dog owners, to override some of those rules and to capitalize on the others. We can only do this if the dog feels secure, because, just as we cannot learn when we are jittery and unsure of ourselves, neither can the dog. A laid-back attitude is therefore very important on the part of the owner. *Oh dear, a mistake on the rug. This isn't good. Not to worry, though, you're smart, you'll get the point. Let's go out and find some places where other dogs are doing it right.* Expressions to this effect should be heartfelt, however, as wherever emotional issues are involved, a dog can usually see through pretense.

The advice to relax applies to all behavioral problems. A generous nature, a willingness to learn what your dog may be thinking, and calmness in the face of a torn shoe or an angry neighbor will go very far toward improvement, especially if the dog believes that you are on his side.

My third bit of advice would be to take with a grain of salt the very prevalent notion that your dog must at all times and at all costs be totally dominated by you. Almost every dog has a perfect understanding of his household's hierarchy and he knows he is not and never will be the dominant member. Nor does he necessarily want to be. Membership in a group is what is important to dogs, not leadership or alpha status, and because the process of domestication has shaped dogs to resemble young wolves, not adult wolves, dogs not only will accept a subordinate position, but want it. A juvenile wolf, alone without leadership, is a very sorry creature, and so are most leaderless dogs.

For some reason, however, many behavioral manuals and how-to books insist that we enforce our dominance. We must be sure to go through a door first, for instance, to make the dog acknowledge our superior status. Nonsense. Of course, we don't want the dog rushing through the door past us, knocking us off our feet, and dogs should be encouraged not to do this, but the symbolism in itself is unimportant. Many a good border collie will always let you go through the door first. In fact, she may insist that you go first by firmly standing back until you are inside. However, she is not deferring to you. Why not? Because she knows all about herding. She wants everybody through that door before she goes, so that she can round up the stragglers.

In fact, I can think of no written advice about enforcing one's dominance that makes much sense. Konrad Lorenz, the great pioneer of animal behavior, describes how he disciplined a dog by picking it up by the back of the neck and shaking it—to be sure, a far more merciful way of punishing a dog than beating it. His purpose was to make the dog feel as if he were being shaken by a much larger dog. A dear friend of mine tried this with a husky and got bitten. In this case, the bite reformed the owner, who never shook the dog again. The shaking did not reform the dog and the outcome merely caused my friend to discredit Lorenz.

I believe that the needless insistence on dominating dogs springs from our society's dog fascism, and the very nature of our relationship with dogs inspires this. Dogs are slaves, whether we like it or not. We buy them and sell them, we legally kill them for reasons that no dog would understand, and we control their reproduction by removing their wombs or testicles or else by choosing partners for them, then taking away their children at an early age. (Most cats, incidentally, still choose their own mates, which is why cats don't suffer from the many ills that afflict dogs, conditions such as deafness, hip dysplasia, and premature aging, to name but three of the many sorry results of our mismanagement.)

We even control the mobility of dogs. We keep them on leashes or in crates. Most communities insist that dogs never run free, even if the dog will not be a nuisance and can do so safely. Dogs cannot run free in most parks, even if the owners pick up the feces. Nor can dogs run free in most wild areas. On a mountainside near my house, on a huge tract of land owned by the state as a wilderness area, dogs must be kept on leashes on the trails. I have known this area since my childhood—in fact, much of it was donated by my father—and in all those years, nothing about the land has changed except that the trees have grown taller and these dog laws have been enacted. Thus the iron hand of dog fascism has reached even to these remote parts.

Even so, I don't put leashes on my dogs when we travel over this mountain. Nobody did so in the past, and I'm not starting now. My dogs are capable and autonomous—they don't harass wildlife, they stay near me, and they come to me when called. But as the dog fascists see it, there is no such thing as a capable or autonomous dog. And these days, that's partly true. A dog who isn't free to learn dog wisdom, either from personal experience or from other dogs, will not be capable or autonomous.

According to dog fascism, there is also no such a thing as a good mongrel. The very existence of a mongrel means that someone got to choose her own mate. Even the word "mongrel" is pejorative, and the euphemisms for mongrels are satirical—"Heinz 57 varieties," for instance. Tucked into the 160 gallons of mail from readers were not a few letters berating me for letting a dog choose a mate. (As mentioned above, the book in question was about the natural behavior of dogs, the choice of a mate being part of this behavior.) This, the readers felt, was "uncontrolled breeding," meaning that I, the owner, did not control the breeding. For a time that was perfectly true. Until I had her spayed, the dog in question rejected several suitors and mated only with the dog she loved, hence *she* controlled her breeding. As the same book describes at some length, however, I did control the fate of the pups, who either remained with us or with our friends for the rest of their lives so that none became homeless or wound up on the streets. Thus I was fully as careful as any breeder as to the pups' well-being, and much more careful than some breeders, and virtually all pet stores. So what? cried the dog fascists. The pups were mongrels!

Yes, they were. I was looking for brains and ability in the pups, not fancy looks. Even so, I couldn't help but notice that breeders of pedigreed dogs do not get accused of "uncontrolled breeding" for the very simple reason that *they,* the human beings, choose the mates of the dogs, and the results are purebred, not mongrel. The notion of racial purity is not of course confined to dogs. Yet it is destructive wherever one finds it.

Nowhere is the mania for rigidly controlling dogs more

clearly shown than in the current fad for castrating male dogs. Many humane societies will not give a male dog for adoption unless the potential owner signs a legally binding contract to guarantee that the dog be castrated. Better that the dog should stay in the shelter, fail to find a home, and be killed. The owner who defaults can be prosecuted.

Of course, every unwanted puppy is a potential tragedy. Nobody argues with that. But castrating male dogs is a bizarre solution. Spaying female dogs is a different story. The only practical way to prevent female dogs from having unwanted puppies is by spaying them. A dog in heat is a powerful sex magnet, and unless she is very carefully monitored, she will almost surely get bred, if not by the neighbor's dog, then by a coyote. Yet spaying a female does not change her profoundly. A female dog is sexually active for just a few days each year. For the rest of the year, her hormones are more or less quiescent and in some ways she might as well be spayed.

Why then, the mania for castration? Some feminists insist that castrating males is only fair. If you're going to spay the females, then castrate the males to keep things equal. Now, I happen to be a feminist myself, but dogs are not, and what is fair for one species is not necessarily fair for all species.

Some veterinarians point out that castration protects dogs from testicular cancers, which may very well be true, but then castration would also protect men, who also get testicular cancers, and we don't see any rush of men hurrying to get their testicles removed, not even the men who insist upon castrating dogs.

And finally, some people feel that castrated male dogs are better behaved than intact male dogs. This too is only minimally true. A very aggressive male might be gentled somewhat by castration, but most male dogs are already tractable, and as for those who aren't, many of the causes of aggression are found in the dog's mind, not in his testicles, in which case castration is useless. A male who roams extensively might also be helped by castration, the key word being *might,* since again, the causes of roaming are usually in the mind. Contrary to a commonly held belief, most intact male dogs who roam are not in search of sex. They'll take it if they find it, certainly, and to that end they'll also follow the tracks of feet with the scent of estrus on them, from stepping on an estrus female's urine, but the random, generalized search for possibly fertile females isn't part of a male dog's cosmos.

Castration can and does change a male dog's personality, but not always in the ways the owner desires, and often the changes are for the worse. Castrated male dogs often seem immature or wimpy, and often have a sad, slow, low-key affect. Because of their odor, and perhaps also because of their behavior, other dogs don't know what to make of them. Male dogs may mount them and female dogs may challenge them and even attack them, and the castrated males are at a loss for what to do about these bizarre reactions.

Even so, every unwanted puppy has a father, and if unwanted reproduction is a problem, then certainly something should be done. But why castration? Why not vasectomy? Vasectomy is fully as effective, much less invasive, much less expensive, and doesn't change the personality of the dog. Any veterinarian should be able to perform the operation (although he certainly shouldn't charge as much for it) because, in fact, almost anyone can perform it. This is by no means to say that just anyone should try it, but it is done in parts of the country where wildlife biologists are attempting to control wolf populations. Even the graduate students of the wildlife biologists can perform the operation in the field. They dart the wolf to tranquilize him, and quickly do the vasectomy while the wolf sleepily looks on. A few minutes later the wolf gets up and leaves. He still has testosterone and the energy it gives him, he still has his personality, he can still keep his place in his pack, he can still stand off other wolves, and he can still penetrate his unspayed mate when she next comes into heat. She won't go looking for somebody else because as far as she's concerned, she's been bred, and by the mate of her choice too. The only difference will be that the vasectomized wolf won't increase the wolf population the following spring, and fewer wolves will overrun the boundaries of the wilderness areas to plague the ranchers. What could possibly be better than that? In theory (and if not for coyotes or for the occasional exceptions made by the dog family to its own marital rules) the entire dog population of the nation could be managed as this wolf population is managed—by vasectomy and female choice alone—no spaying, no castration, no unwanted pups. Yet dogs are not so fortunate, and in the face of vasectomy, the dog fascists continue to promote castration.

Rigid, unthinking control is not a good solution to any question concerning animals. We'd all be better off if dog fascism were controlled.

In the 160 gallons of mail were also many letters from women whose husbands or boyfriends didn't like their dogs. In every case, the man wanted the woman to get rid of her dog. However the woman loved the dog and was placed in a terrible dilemma, hence the letter. My advice on this question is always the same. Lose the man. Keep the dog. You are far better off with the dog than with a man who would ask such a thing of you.

Appendix II

A Note on Keeping Parrots

BIRDS are very difficult pets indeed, and parrots are among the most difficult. Two parrots are four times as much trouble as one, because the level of difficulty rises exponentially. So can the level of disappointment. Many people buy parrots expecting them to talk, but many of the so-called talking parrots won't say a word. Instead, they make sounds like the microwave or the dishwasher, or the phone ringing, or a person coughing, or the dog barking, or, worst of all, the smoke alarm. And if you are lucky enough to find a parrot who does talk, it won't say much unless you spend many hours talking with it, or at the very least, talking with someone else in its presence. The tape recordings and other devices which are sold to teach speech to parrots simply do not work, because parrots vocalize in order to communicate, as they would in the wild. The microwave, the smoke alarm, and other household devices *do* communicate, although we may not think of their noises as such, yet when we hear them, we respond, as the parrots notice.

Even if you buy a parrot who is known as a talker, the bird may still decline to speak. One of my parrots (my favorite, actually) knows dozens of words but almost never uses them. Quite literally, months can pass before she says anything. If I had purchased her as a talker, which she is, I would have been bitterly disappointed at how seldom she uses the ability.

Don't even think about getting a bird unless you are ready to put up with a huge mess, as any bird will scatter food for long distances and will also shed feathers every day, which will blow around the room. Some kinds of parrots also make dust, as they grow a special kind of feather especially for that purpose. When they preen, the powder from these feathers conditions their other feathers, but it also flies around the house, coating surfaces just as pollen does, and making people cough. And then there's the cage, which must be kept scrubbed clean and supplied with fresh paper at frequent and regular intervals or the bird will get sick.

A parrot will also get lonely and bored unless you are prepared to spend a considerable amount of time with it. If you leave home for any length of time, you must either take it with you or make arrangements for its care, not the best or cheapest of which is boarding it. Unless yours is the only bird at the boarding facility, it may catch a disease from some other boarder. On the other hand, if it is the only bird, it will probably feel very lonely and may stop eating or worse. But never think that you can easily provide a companion by buying another bird, because birds are not like dogs and cannot be expected to bond with one another, certainly not if they are of the same sex. However, unless their DNA has been studied (very expensive), only they will know this. Parrots unfailingly perceive the gender of human beings, but because most species of parrots do not have sexual dimorphism, we cannot do the same for them, and the bird breeders and pet stores virtually never do the DNA studies before putting the birds up for sale. The normal procedure is to buy a bird, and *then* discover its sex, assuming you're prepared to spend the money. Alas, same-sex parrots are often bitter rivals, but breeders and pet stores rarely take a bird back for any reason, certainly not because the gender is wrong.

Few pets are more expensive than a parrot. A pair of hyacinth macaws—not a breeding pair or a mated pair, just two hyacinth macaws—sold recently for $15,000 each. I don't claim to know much about bird prices, but $2,000 or $2,500 would not be a surprising price these days for a young, healthy parrot or cockatoo or macaw—maybe more. Naturally, such prices attract thieves, which puts your bird at continuous risk of being stolen.

Still, one does well to pay the prices asked by reputable breeders, since the less expensive parrots found elsewhere often have histories of which the purchaser is unaware, and only reputable breeders can be depended upon to be truthful. An avian veterinarian tells the cautionary story of a tourist to Central America who bought a cockatoo from a vendor on the street. The vendor told the tourist that the bird was young and had been caught locally, and also that it would not require a cage, because a T-stand perch would be sufficient. In fact, the bird was not young and had not been caught locally, as cockatoos are indigenous to Australia and Indonesia and the islands in between, not to Central America. But the tourist didn't know any of this.

He took the bird home and put it on a T-stand, and when he returned from work the following day he found that his house had been trashed. Free to do as it pleased, the bird had torn his books, shredded his curtains, and taken bites out of his furniture, to say nothing of splashing big white droppings all over everything. The distraught man hurried to the nearest store that sold bird supplies, bought a large, ruinously expensive cage and put the bird inside. But the bird had not been caged before, and didn't like it, and by the time the man got home from work the following day, the bird was naked. It had plucked out its feathers.

Actually, there are two cautionary aspects to this story, the first being that a bird is not an object, but instead has memory and emotions, hence its past and also its preferences will be important and should be known. The second cautionary aspect is that before a bird is purchased, it should be seen by an avian veterinarian, as was the naked cockatoo the next morning. The veterinarian who saw the bird and later told the story helped solve its problems by finding it a new home.

But there are additional reasons to see an avian vet rather than a vet who normally treats just dogs and cats, not the least of which is that the difference between a bird and a mammal is so great. Yes, birds are "animals," as are dogs and cats, yet dogs and cats are much closer to human beings than they are to birds, and no veterinarian trained to treat dogs and cats would be allowed to treat a person. The same veterinarian should also be reluctant to treat a bird. Meanwhile, avian veterinary medicine is so specialized that these vets are few and far between. I bring my birds to an avian vet who is almost fifty miles away, making a round trip of a hundred miles for every visit, and I'm extremely lucky that she's so near. She is also an excellent doctor. The same cannot be said for every avian vet, hence any prospective bird owner should do the necessary research to find a good one.

Even so, an avian practice requires not only highly specialized knowledge, but also specialized equipment and testing procedures, none of which come cheap. I spent upwards of $1,500 to treat my macaw, who turned out to have a disease for which there is no cure (very little is known about many bird diseases) and which always results in death. I took the bird to an excellent veterinarian, but even then, what did I get after spending enough money to buy a new winter wardrobe or take a vacation? One dead bird—yet one whom I loved with all my heart, and would have sold the house if money could have saved her.

Must one take one's parrots to a vet? Yes, indeed. Parrots need to have their claws trimmed, their wings trimmed, and sometimes even their beaks trimmed. Parrots cannot be immunized, so their droppings and blood must be checked at intervals for conditions that can lead to serious symptoms. Many bird diseases are as silent as they are contagious.

Bird food is also expensive. You cannot give your bird a handful of seeds from the local supermarket and expect it to thrive. While the dietary requirements are different for every kind of bird, parrots need a breakfast that includes a smorgasbord of fresh fruit and vegetables, also a little bread or pasta, sometimes a little chopped hard-boiled egg, or perhaps a little cheese. A good dinner might include a helping of some respected brand of bird kibbles and a helping of some commercial mixture of dry fruit, dry vegetables, seeds, and nuts. These items are specialized and very expensive, but not all brands are equally nutritious, and none are readily available. You will have to do research to find the appropriate kinds, and you will probably have to travel to buy them, or order them by mail in bulk, then keep them in the freezer because they will not have been treated with insecticide (poison to birds) and will be harboring moths and their larvae. The evening meal might also include a few peanuts, a dried red pepper or two, some raisins, and perhaps a few pine nuts or some other kind of nut—one that the bird can open. But the above is merely an example. Any prospective parrot owner cannot be too careful in learning the particulars of his own bird's needs.

A note here: Birds are highly sensitive to bacteria, so any fresh food must be removed from the cage quite quickly. Certain foods are poison to birds, most notably chocolate and avocado. But then, many things are poison to birds (including household fumes, especially those from Teflon). In the wild, birds learn about food from their elders, and hence are likely to eat whatever their owners give them—in other words, a bird cannot be depended upon to know what's good for it and what is not. If loose, birds will also forage on their own, and will eat the poison leaves and berries of common houseplants if these are available.

As for the cage, the bigger the cage the more it costs, yet expense cannot be spared in getting a cage with proper volume. At the very least, a cage should be big enough for the bird to spread its wings in any direction, and bigger if possible. Anything smaller becomes a straitjacket, and is the ultimate in cruelty to an animal whose natural habitat is the sky. Inside the cage, the bird should have—among other items—perches suitable to its feet and a supply of specially made bird toys.

Too much cannot be said for proper toys. Parrots are frighteningly intelligent—the nickname "birdbrain" is the world's greatest misnomer—and without something to play with they get bored and miserable, which can cause them to stop eating or to pluck out their feathers. Since birds are uncommon pets (as you probably realize by this time) the manufacturers of bird toys cannot sell their product in volume so must charge astronomical prices. A price tag of $16 to $20 for a toy about eight inches long would not be unusual. A bird should have several toys at a time, and new toys should be added fairly often or the bird will get bored and scream tediously or pluck out its feathers. Yet pay you must, unless you make your own bird toys, which requires considerable knowledge. You cannot, of course, use materials that will harm the bird, and research is required to determine what these materials might be.

No discussion of the downside of bird ownership should fail to include noise and destruction. Parrots excel at making noise. Many of them live in thick forests, yet communicate by voice, hence their calls are designed to travel very far in unfavorable conditions. The screams of my cockatoo can be heard a mile away, quite literally. She's so loud she gives people headaches. Some people even get toothaches. Nobody but me will stay in my office while she's screaming. She screams when she hears me speaking (she wants to join in), hence she screams when I'm talking on the phone, and the racket she makes is so piercing that callers get confused and can't remember what they are saying. Nor can they hear what I am saying. I often must go to another phone elsewhere and call them back.

Virtually all parrots scream. My mealy Amazon parrot screams hideously for no real reason, unpredictably, and for hours on end. She had been given up by her former owners, perhaps because she deafened them. Covering her cage does no good—she screams anyway. She can only be stopped at night, and only by turning out the lights. She's very sweet, though—she screams until the lights go off, then she softly says, "Oh?" and falls silent. And as for the rest of us, we must continue our conversations in pitch darkness—or move to another room.

All parrots are destructive. Many of them make their nests in hollow trees, hence they know how to take big bites out of wood, and unfailingly ruin woodwork and furniture when released from their cages. Yet they must be let out from time to time, as without a little freedom, they get bored and scream or pluck out their feathers. I don't believe I own a book that hasn't been partly destroyed by a parrot. A parrot can bite through a broomstick, and if she bites you, which sooner or later she will, you'll know it. Remember those pictures of a pirate with a patch over his eye and a parrot riding on his shoulder? Could the parrot have had anything to do with the loss of the eye? Parrots can bite through electric cords, shorting out the entire house and killing themselves in the process, and they unobtrusively chew holes in your clothing while they are sitting on the back of your chair. They also bite your dogs and cats if these animals unwisely go near them. We had four cats at the time I got my first parrot, the mealy Amazon, and our friends were afraid that the cats would reach through the bars of her cage and harm her. But I saw the problem differently. I was afraid I'd end up with one parrot and four three-legged cats. I needn't have worried. The cats were way too smart to put their paws into the cage with that green demon.

Finally, many birds, especially parrots, normally live for about sixty years but can easily live for eighty years or longer, or in other words, anyone old enough to have earned the money to buy a parrot will probably predecease it. Hence one's commitment to a parrot extends beyond the grave.

Is having a parrot worth all this? Most certainly. But no one should get a parrot without a clear understanding of the staggering amount of difficulty that owning such a bird involves.

Acknowledgments

My deepest thanks go to the animals mentioned in the dedication of this book, and to our beloved Sundog in particular. Without him, this book would not have been written. I would also like to thank the people who participated in our household group, mentioning first the people who were most often present: my husband, Steve; my mother, Lorna Marshall; our dear friends, Sy Montgomery, Donald Schrock, Susan Culver, Vladimir Pistalo, Peter Schweitzer, Eleanor Peron, Howard Mansfield, Anna Martin, Rebecca Adams, Janice Chalke, and Carol Lambert. I would also like to thank the members of our family who could not be present as much as we would have liked: our daughter, Stephanie, and her husband, Bob Kafka; our son, Ramsay; and our grandchildren, David, Zöe, Ariel, and Margaret. How deeply they belonged among us was made clear by Sundog, who never forgot them.

I am also very grateful to my agent, John Taylor Williams, of the Palmer Dodge Agency in Boston, to Jared Williams, who provided the illustrations, and to my publishers as well.

Those of us who share our lives with animals greatly appreciate the veterinarians who care for them, and to this end I would like to thank the doctors and staff of two splendid veterinary practices, the Animal Care Clinic in Peterborough, New Hampshire, and the Littleton Animal Hospital in Littleton, Massachusetts, especially Dr. Charles DeVinne, Dr. Janet Mack, Dr. Alexandra Kilgore, Nancy Lobacki, Nancy Traffie, Sandy Boutwell, Judy Traffie, and Kelly Lambert.

About the Author

Elizabeth Marshall Thomas is also the author of The Tribe of Tiger, The Harmless People, Warrior Herdsmen, The Hidden Life of Dogs, *and the novels* Certain Poor Shepherds, Reindeer Moon, *and* The Animal Wife. *She lives in Peterborough, New Hampshire.*